Complete Guide

to High School Math

MASTER PRECALCULUS

WITH GEOMETRY & ALGEBRA REVIEW

WRITTEN BY JOE SUNG

Edited by Demi_Oh
Designed by Demi_Oh
Illustrated by Demi_Oh

PRECALCULUS

ⓒ Joe Sung, 2021

초판 1쇄 발행 2021년 3월 19일

지은이   Joe Sung(성혁진)
펴낸이   이기봉
표지/내지 디자인 Demi Oh(오혜경)
편집   Demi Oh(오혜경), 좋은땅 편집팀
펴낸곳   도서출판 좋은땅
주소   서울 마포구 성지길 25 보광빌딩 2층
전화   02)374-8616~7
팩스   02)374-8614
이메일   gworldbook@naver.com
홈페이지   www.g-world.co.kr

ISBN   979-11-6649-464-2 (53410)

DEDICATED TO

My lovely wife Demi Oh

My precious son & daughter Ryan & Skylar Sung

My devoted parents Peter & Susana Sung

My thankful mother-in-law Sook Hyun Ki

My reliable brother Robin Sung

WRITTEN BY JOE SUNG

Edited by Demi_Oh
Designed by Demi_Oh
Illustrated by Demi_Oh

Preface

Author Bibliography

" 탄탄한 개념+확실한 고득점 노하우 "

Joe Sung 선생님

- 현 블루리본에듀학원 Math 대표강사
- 전 ESC, 삼보어학원 Math 대표강사
- 전 Sophis 어학원 원장
- 전 명덕, 이화, 고양외고 Math 대표강사
- 캐나다 고교 졸
- 캐나다 UBC Pre-Med. 전공 졸
- Math 16년 이상 강의 경력
- 전 과목 영어+한국어 강의!!

JOE SUNG / 성 혁 진

저자가 한국에 와서 학생들에게 수학을 가르치기 시작한지 어느새 20년이 가까운 시간이 흘렀다. 그 동안 본 저자의 제자들은 대부분 IVY League를 졸업하고 다양한 분야에서 본인의 이름을 빛내고 있다. 실로 자랑스러운 일이 아닐 수 없다.

그 긴 시간동안 수학 시험은 다양한 형태로 변화했지만 결국 그 안의 학문에 대한 핵심은 늘 그대로이다. 이에 저자는 학생들이 더 이상 불필요한 학습으로 시간을 낭비하지 않고 단기간에 효과적으로 학생 본인이 원하는 목표에 다다를 수 있도록 도울 수 있는 방법을 고민하게 되었다.

그리고 그 고민은 미뤄두었던 이 책의 발간을 시작으로 그 첫 발을 내딛고자 한다.

오랜 기간 고민하고 작업하여 이루어진 결실인만큼 학생들이 본 교재를 효과적으로 사용해주기 바라는 마음이다. 다음 책은 최대한 빠른 기일내에 추가 발간될 예정이며 앞으로도 학생들의 효과적이고 확실한 Math 학습을 위해서 고군분투할 것이다.

Contact Info.: joesung1@gmail.com / Instargram ID : joesung_math

저자가 말하는 바른 공부를 위한 Guideline

저자가 직강을 할 때 그리고 설명회를 할 때
늘 강조해서 이야기하는 것이 있다.
그 내용을 이 책에서도 함께 나누려고 한다.
저자가 제시하는 5가지 GUIDELINE에 맞춰
효과적인 MATH학습을 해보자

1. 효과적인 MATH의 공부순서
가능하다면 아래의 순서대로 공부하자

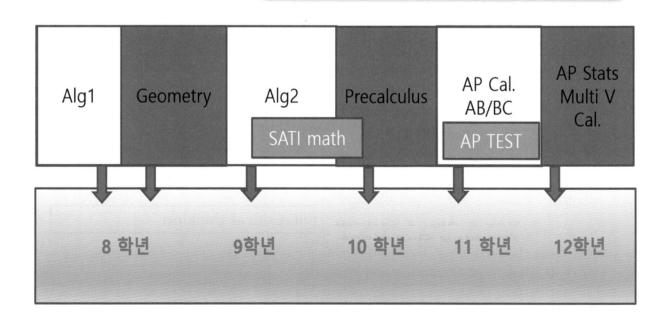

2. CUSTOMIZED MATERIALS
시험 및 성적에 도움이 되는 문제로 공부하자

16년 동안의 경력을 바탕으로 한 다양하고 정확한 Joe샘 개인자료

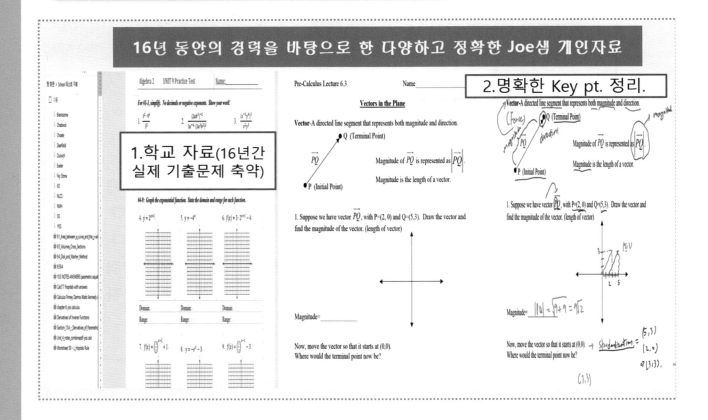

2.명확한 Key pt. 정리.

1.학교 자료(16년간 실제 기출문제 축약)

3. 확실한 CONCEPT / 실력향상문제
배운 건 잊지않도록, 실력이 늘 수 있는 문제로 공부하자

i. 문제풀이는 항상 accumulative한 문제로 꾸준히

ii. connection

iii. Critical thinking

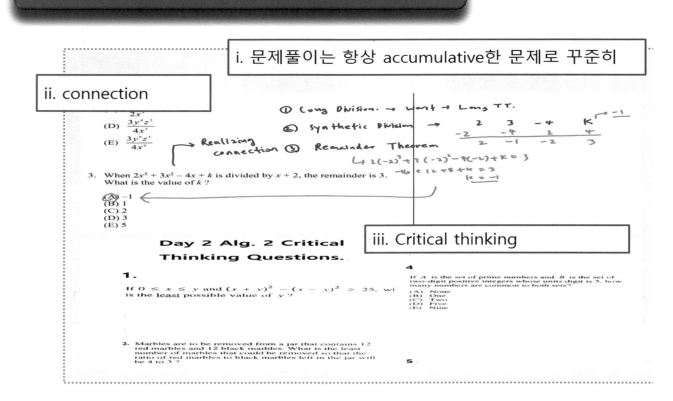

4. 오답 노트 / MENTORING
오답노트는 정확하고 꼼꼼하게 다시 복습할 수 있도록 하자

✓ 학생에게 맞는 **Template**제공

✓ 의지력 향상 ✓ 동기 부여 ✓ **Communication**

✓ 꾸준히 공부할 수 있게 늘 도와주는 **Mentor**

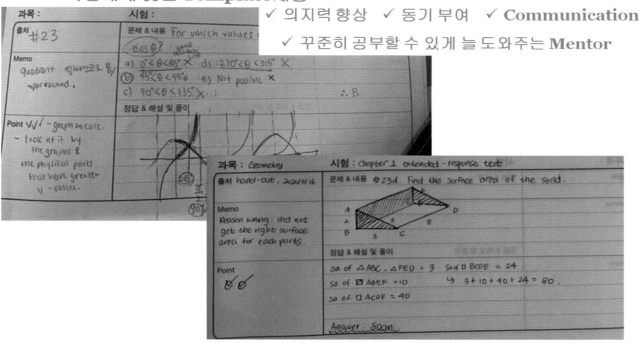

5. 바른 습관
스스로의 의지로 아래의 습관을 만들자

포기하지 않는 습관

힘들거나 모르는 문제가 나왔을 때도 포기하지 않고 집요하게 풀어내려고 노력하는 습관.

완벽을 추구하는 습관

틀린 문제를 그냥 넘어가지 않고 꼼꼼하게 오답노트를 하면서 다음에는 꼭 맞추려고 하는 완벽을 추구하는 습관.

도전하는 습관

현재 단계에서 쉽게 만족하고 안주하지 않고 다음 단계를 향해 준비하고 도전하는 습관.

성취하는 습관

자신의 Goal을 확실하게 설정하고 성취해내는 습관.

Preface

Structures and Features

1. 본 교재는 본 저자의 강의를 듣는 학생이 저자의 수업을 들으면서 보다 효과적으로 학습하고 Precalculus과목을 확실하게 Master할 수 있도록 최상의 자료를 제공하는 것을 목표로 합니다.

2. 본 교재는 국내 외국인 학교 혹은 외국 Boarding School에 다니고 있는 학생이 스스로 학교 수학을 완벽하게 Master할 수 있도록 학교 Text Book과 최적의 조화를 이룰 수 있는 문제를 제공하는 것을 목표로 합니다.

3. 본 교재는 Precalculus 전반에 걸친 핵심 Concept을 이해하고 이에 적합한 문제들을 제공하는 것뿐 아니라 Algebra2와 Geometry Review Section을 통해 복습할 수 있도록 함으로써 미국 High School 수학의 기본을 다질 수 있도록 만들어졌습니다.

4. 본 교재는 국내 외국인 학교 혹은 외국 Boarding School에서 많이 쓰는 교재를 중심으로 편집되어 있어 불필요한 문제들을 최소화하였고 학교에서 많이 다루는 개념을 중점으로 Concept에 대한 정리가 확실하게 되어있어 학교 내신 시험 준비에 실질적인 도움이 될 수 있도록 만들어져 있습니다.

5. 본 교재는 수학을 한 번에 끝낼 수 있도록 고안되었으므로 완벽한 연습과 스스로의 노력을 통해 오답률 0%를 만드는 것을 목표로 합니다.

Preface

1. This textbook aims to allow students to learn more effectively and master Precalculus by following along the author's lectures and textbook as their study guide.

2. This textbook contains problems that can optimally be harmonized with school textbooks in order to allow students attending a foreign school in Korea or a boarding school in the U.S. to master Precalculus on their own.

3. This textbook is designed to strengthen the basics of high school mathematics in the United States by providing understandings of core concepts of the entire Precalculus, followed with appropriate practice problems, as well as review materials for Algebra2 and Geometry.

4. This textbook is edited to match textbooks that are often used in foreign schools in Korea and boarding schools in the U.S. in order to minimize unnecessary concepts.

5. This textbook is designed to complete math at once, so we aim to make a 0% inaccuracy rate through practice and self-effort.

Preface

How to use it

1. Precalculus를 공부하기에 앞서 목차를 참고하여 뒷부분의 Algebra2와 Geometry Review Section을 통해 본인의 수준을 체크해보도록 합니다.

2. 복습이 다 이루어지고 나면 학교에서 공부하고 있는 Precalculus textbook의 목차와 본 교재의 목차를 비교하여 순서대로 차근차근 공부하여도 좋고 본인 스스로 약하다고 생각하는 부분이나 공부하고 싶은 부분의 Note section의 문제들을 풀어봅니다. 문제를 풀기 전 해당하는 문제의 Formula나 Concept에 대한 이해를 확실히 합니다.

3. 전체적인 Precalculus를 완벽히 공부하고 나면 Note section을 이용해서 전체적인 section을 스스로 복습하고 Worksheet에서 전에 공부한 부분을 추가적으로 문제풀이를 할 수 있도록 되어 있으므로 Note section을 스스로 공부한 후에는 Worksheet의 추가적인 문제들을 꼼꼼하게 풀도록 합니다.

4. Note와 Worksheet의 모든 문제들은 반드시 오답노트를 만들어 다시 복습할 수 있도록 합니다. 이는 수학 공부의 아주 중요한 부분이므로 절대 간과하지 않도록 합니다.

5. 마지막으로 본 교재를 가장 효과적으로 사용하고 Precalculus를 가장 효과적으로 Master하는 방법은 한 번 문제를 풀 때 정확하게 풀고 틀린 문제들을 오답노트하여 본인이 약한 부분에 대한 확실한 자가 진단을 하고 그 부분을 강화시키는 것입니다.

모든 공부는 한번에 끝내려는 습관을 갖는 것이 중요합니다. 두 번 세 번 공부하는 파트는 반드시 눈에 띄게 줄어 결국에는 오답 0%가 될 수 있도록 합니다.

How to use it

1. Before studying Precalculus, refer to the Table of Contents, and check your level through Algebra2 and Geometry Review Section at the end.

2. After the reviewing, you may compare Table of Contents of the Precalculus textbook you are using at school with that of this textbook to study correspondingly. Before solving a problem, make sure you understand the formula or concept of the problem.

3. Upon completion of Precalculus contents, you should review the Note section to further solidify your knowledge. You should also solve problems in the Worksheet section to thoroughly examine additional problems in the Worksheet.

4. For all questions in the Note and Worksheet sections, making yourself review notes for wrong answers will help you revisit each incorrect problem. This is a very crucial part of studying math, so don't overlook it.

5. Lastly, the most effective way to use this textbook to master Precalculus is solving problems correctly and self-diagnosing yourself on your weaknesses. You can reinforce your weaknesses by accurately solving the problem again and making review notes for wrong answers.

It is important to get in the habit of completing all your studies at once. The content that you miss and study two or three times must be noticeably reduced to 0% inaccuracy rate.

Table of Contents

Table of Contents

WorkSheet

Table of Contents

Table of Contents

Geometry

Algebra2

Formulas for PreCalculus

Memorizing formulas is the most necessary step
in studying high school math.   -Joe Sung.

Formulas

Formulas for PreCalculus

The mathematical formulas that you must
memorize before entering AP Calculus.
(Precalculus, Algebra2, and Geometry formulas)

1) Function Notation and Inverse property.

$(f \circ g)(x) = f\big(g(x)\big)$

If $(f \circ g)(x) = f\big(g(x)\big) = x$, then f(x) is an inverse function of g(x)

If f(g(x)) = g(f(x)), f(x) is equal to g(x) or f(x) is inverse function of g(x)

2) Linear Function

$Ax + By + C = 0$:   standard form of line.

a) The shortest distance between a line and a point $= \dfrac{|Ax+By+C|}{\sqrt{(A^2+B^2)}}$

(x, y) = xy coordinate point

A, B, C are coefficients of a line

b) The acute angle, θ between two line: $\operatorname{Tan}\theta = \left|\dfrac{m_1-m_2}{1+m_1 m_2}\right|$

$m_1 =$ slope of a line 1

$m_2 =$ slope of a line 2

c) Basic formulas

i) Distance d between 2 points, $P(x_1, y_1)$ and $Q(x_2, y_2)$:

$d(P, Q) = \sqrt{(x_2 - x_1)^2 + (y_2 - y_1)^2}$

ii) Distance d between 2 points in 3Dimension (x_1, y_1, z_1) and $(x_2, y_2, z_2) =$

$$\sqrt{(x_2 - x_1)^2 + (y_2 - y_1)^2 + (z_2 - z_1)^2}$$

iii) Mid-point: $\left(\dfrac{x_1+x_2}{2}, \dfrac{y_1+y_2}{2}\right)$

iv) Slope of a line: $m = \dfrac{y_2-y_1}{x_2-x_2}$

v) Condition for parallel lines: $m_1 = m_2$

vi) Condition for perpendicular lines: $m_2 = \dfrac{-1}{m_1}$

If multiple of two slopes are -1, both lines are perpendicular.

$(m_1 \bullet m_2 = -1)$

d) Equation of a Line

The point-slope form, slope m and goes through (x_1, y_1): $y - y_1 = m(x - x_1)$

The slope-intercept form, slope m and y-intercept b : $y = mx + b$

3) Quadratic Functions

$y = ax^2 + bx + c$: standard form

a) Vertex $= \left(-\dfrac{b}{2a}, c - \dfrac{b^2}{4a}\right)$

i) $-\dfrac{b}{2a} =$ Axis of Symmetry

ii) $c - \dfrac{b^2}{4a} =$ Min if a $= +$

                   Max if a $= -$

b.) Quadratic Formula

If $a \neq 0$, the solution of the equation $ax^2 + bx + c = 0$ are given by $x = \dfrac{-b \pm \sqrt{b^2 - 4ac}}{2a}$

c) Discriminant Law (for quadratic equation)
i) $b^2 - 4ac > 0$: there are 2 real solution (2 x-intercept)
ii) $b^2 - 4ac < 0$: there are 2 non-real solution (no x-intercept)
iii) $b^2 - 4ac = 0$: there is one distinct real solution. (double roots) (1 x-intercept)

d) Relationship between coefficients and roots
$ax^2 + bx + c = 0$
i) sum of roots $= \dfrac{-b}{a}$
ii) product of roots $= \dfrac{c}{a}$

4) Even and Odd function

a) Even function

i) $f(x) = f(-x)$
ii) symmetric about y-axis.
iii) For polynomial: if all exponents are even, the function is even.

b) Odd function

i) $f(x) = -f(-x)$
ii) symmetric about the origin
iii) For polynomial: if all exponents are odd and the function has no constant, the function is odd.

5) Formulas from Trigonometry

[Angular Measure] π radians $= 180°$

So, 1 radian $= \dfrac{180}{\pi}$ degrees,

and 1 degree $= \dfrac{\pi}{180}$ radians

a) To find radian from degree measure: $\theta_r = \theta° \times \dfrac{\pi}{180°}$

b) To find degree from radian measure: $\theta° = \theta_r \times \dfrac{180°}{\pi}$

6) Trigonometry Identities

a) Definition.

$\sin x = \dfrac{1}{\csc x}$ $\csc x = \dfrac{1}{\sin x}$

$\cos x = \dfrac{1}{\sec x}$ $\sec x = \dfrac{1}{\cos x}$

$\tan x = \dfrac{\sin x}{\cos x}$ $\cot x = \dfrac{1}{\tan x}$

b) Special relation.

$\sin(-x) = -\sin x$, $\csc(-x) = -\csc x$: odd relation.

$\cos(-x) = \cos x$, $\sec(-x) = \sec x$: even relation

$\tan(-x) = -\tan x$, $\cot(-x) = -\cot x$: odd relation

c) Important Trig. Identities for standardized test.

1. $\sin^2 x + \cos^2 x = 1$
2. $\tan^2 x + 1 = \sec^2 x$
3. $1 + \cot^2 x = \csc^2 x$
4. $\sin(u + v) = \sin u \cos v + \cos u \sin v$
5. $\sin(u - v) = \sin u \cos v - \cos u \sin v$
6. $\cos(u + v) = \cos u \cos v - \sin u \sin v$
7. $\cos(u - v) = \cos u \cos v + \sin u \sin v$
8. $\tan(u + v) = \dfrac{\tan u + \tan v}{1 - \tan u \tan v}$
9. $\tan(u - v) = \dfrac{\tan u - \tan v}{1 + \tan u \tan v}$

Double-Angle Identities
10. $\sin 2u = 2 \sin u \cos u$

11. $\cos 2u = \cos^2 u - \sin^2 u$

12. $\qquad = 2\cos^2 u - 1$

13. $\qquad = 1 - 2\sin^2 u$

14. $\tan 2u = \dfrac{2\tan u}{1-\tan^2 u}$

Cofunction Identities

$$\cos\left(\frac{\pi}{2} - u\right) = \sin u$$

$$\sin\left(\frac{\pi}{2} - u\right) = \cos u$$

$$\tan\left(\frac{\pi}{2} - u\right) = \cot u$$

$$\cot\left(\frac{\pi}{2} - u\right) = \tan u$$

$$\sec\left(\frac{\pi}{2} - u\right) = \csc u$$

$$\csc\left(\frac{\pi}{2} - u\right) = \sec u$$

7) Triangles

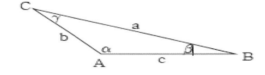

a) Law of sines:

$$\frac{\sin A}{a} = \frac{\sin B}{b} = \frac{\sin C}{c}$$

b) Law of cosines:

$$a^2 = b^2 + c^2 - 2bc \cos A$$
$$b^2 = a^2 + c^2 - 2ac \cos B$$
$$c^2 = a^2 + b^2 - 2ab \cos C$$

c) Area $= \frac{1}{2}bc \sin A = \frac{1}{2}ac \sin B = \frac{1}{2}ab \sin C$ (for any triangle)

d) Area $= \sqrt{s(s-a)(s-b)(s-c)}$ Where $s = \frac{1}{2}(a+b+c)$

8) Trig. Graphing

a) sine: $y = A \sin[b(x - c)] + d$ b) cosine: $y = A \cos[b(x - c)] + d$

i) amplitude A,

ii) period $2\pi \div b \implies \frac{2\pi}{b}$

iii) page shift $(x - c) \implies$ (c units right), $(x + c) \implies$ (c units left)

iv) vertical shift d unit up

v) max= d+A

vi) min=d−A

9) Conic Section

a) Equation of a Circle

The circle with center (h, k) and radius r: $(x - h)^2 + (y - k)^2 = r^2$

b) Equation of Ellipse

$$\frac{(x - h)^2}{a^2} + \frac{(y - k)^2}{b^2} = 1$$

i) focal length: If a > b, focal length $= (\sqrt{a^2 - b^2})$

If b > a, focal length $= (\sqrt{b^2 - a^2})$

ii) Length of Major axis = 2a (if a>b, a= always longer stretch)

iii) Length of Minor axis = 2b (if a>b)

iv) Area of ellipse $= \pi ab$

c) Equation of hyperbola

i) The hyperbola with x-stretch a and y-stretch b, and center (h, k):

$$\frac{(x - h)^2}{a^2} - \frac{(y - k)^2}{b^2} = 1$$

ii) transverse axis = 2a since $(x - h)^2$ is positive

d) Two intersecting lines

$(x - h)^2 - (y - k)^2 = 0$ with intersection (h, k)

10) Exponents and radical formulas

a) Exponents:

If all bases are non-zero:

$$u^m u^n = u^{m+n}$$

$$u^0 = 1$$

$$(uv)^m = u^m v^m$$

$$\left(\frac{u}{v}\right)^m = \frac{u^m}{v^m}$$

$$\frac{u^m}{u^n} = u^{m-n}$$

$$u^{-n} = \frac{1}{u^n}$$

$$(u^m)^n = u^{mn}$$

b) Radicals and Rational Exponents

If all roots are real numbers:

$$\sqrt[n]{uv} = \sqrt[n]{u} \cdot \sqrt[n]{v}$$

$$\sqrt[m]{\sqrt[n]{u}} = \sqrt[mn]{u}$$

$$\sqrt[n]{u^m} = (\sqrt[n]{u})^m$$

$$u^{1/n} = \sqrt[n]{u}$$

$$u^{m/n} = (u^m)^{1/n} = \sqrt[n]{u^m}$$

$$\sqrt[n]{\frac{u}{v}} = \frac{\sqrt[n]{u}}{\sqrt[n]{v}} \ (v \neq 0)$$

$$(\sqrt[n]{u})^n = u$$

$$\sqrt[n]{u^n} = \begin{cases} |u|, if \ n = \ even \\ u, \ if \ n = \ odd \end{cases}$$

$$u^{m/n} = (u^{1/n})^m = (\sqrt[n]{u})^m$$

11) Logarithms formulas

If $0 < b \neq 1, 0 < a \neq 1, and \ x, R, S > 0$
$y = \log_b x$ if and only if $b^y = x$

$$\log_b 1 = 0$$

$$\log_b(b^y) = y$$

$$\log_b(RS) = \log_b R + \log_b S$$

$$\log_b R^c = c \log_b R$$

$$\log_b b = 1$$

$$b^{\log_b x} = x$$

$$\log_b \frac{R}{S} = \log_b R - \log_b S$$

$$\log_b x = \frac{\log_a x}{\log_a b}$$

12) Special Products

$$(u + v)^3 = u^3 + 3u^2v + 3uv^2 + v^3$$
$$(u - v)^3 = u^3 - 3u^2v + 3uv^2 - v^3$$
$$u^3 + v^3 = (u + v)(u^2 - uv + u^2)$$
$$u^3 - v^3 = (u - v)(u^2 + uv + u^2)$$

13) Inequalities

$If\ u < v$ and $c > 0, then\ uc < vc$
$If\ u < v$ and $c < 0, then\ uc > vc$
$If\ c > 0, |u| < c$ is equivalent to $-c < u < c$
$If\ c > 0, |u| > c$ is equivalent to $u < -c$ or $u < c$

14) Multiplication of Matrix

$$\begin{bmatrix} a & c \\ b & d \end{bmatrix} \cdot \begin{bmatrix} e & g \\ f & h \end{bmatrix} \Longrightarrow \begin{pmatrix} ae + cf, ag + ch \\ be + df, bg + dh \end{pmatrix}$$

15) Determinants

$$\begin{vmatrix} a b \\ c d \end{vmatrix} = ad - bc$$

$$\begin{vmatrix} a & b & c \\ d & e & f \\ g & h & i \end{vmatrix} = aei + bfg + cdh - ceg - bdi - afh$$

16) Polar and Rectangular conversion

$$(r, \theta) \Longrightarrow r \cos \theta \Longrightarrow x$$
$$r \sin \theta \Longrightarrow y$$
$$r^2 = x^2 + y^2$$
$$\theta = tan^{-1}\left(\frac{a}{b}\right)$$

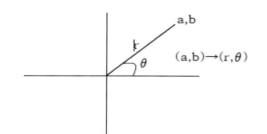

17) Parametric

a) $x = a + bt$
   $y = c + dt$
   -substitute t into one of the equations and find slope, x and y intercepts.
   (slope $\Rightarrow \frac{d}{b}$)

b) need to know how to graph in parametric mode

18) Permutations and Combination

i) Factorial
$n! \Rightarrow n(n-1)(n-2) \cdots 2 \cdot 1$
$(0! = 1)$

ii) Permutation
$_nP_r = \dfrac{n!}{(n-r)!}$

iii) Combination

$_nC_r = \dbinom{n}{r} = \dfrac{n!}{r!(n-r)!} = \dfrac{nPr}{r!}$ (integer n and r, $r, n \geq r \geq 0$)

19) Binomial Theorem

If   n is   a positive integer

$(a+b)^n = \dbinom{n}{0}a^n + \dbinom{n}{1}a^{n-1}b + \cdots + \dbinom{n}{r}a^{n-r}b^r + \cdots + \dbinom{n}{n}b^n$
$(a+b)^n = {}_nC_o\, a^n + {}_nC_1 a^{n-1} \cdot b^1 + {}_nC_2\, a^{n-2} \cdot b^2 \cdots {}_nC_n\, a^0 b^n$

20) Sequences and Series

a) Arithmetic Sequences and Series

i) $a_n = a_1 + (n-1)d$

ii) $S_n = n\left(\frac{a_1+a_n}{2}\right)$ or   iii) $S_n = \frac{n}{2}[2a_1 + (n-1)d]$

b) Geometric Sequences and Series

i) $a_n = a_1 \cdot r^{n-1}$

ii) $S_n = \frac{a_1(1-r^n)}{1-r}\ (r \neq 1)$

c) Infinite geometric series: $S_n = \frac{a_1}{1-r},\ (-1 < r < 1) => only\ when\ this\ satisfies$

21) Domain and Range

a) $y = \dfrac{(x-a)(x-c)}{(x-a)(x-b)}$

> - if denominator $= 0$; $x=a$, $x=b$ (undefined)
>
> Domain {all real numbers except $(x=a, x=b)$}

i) Vertical asymptotes $x=b$

ii) hole $\Rightarrow x = a$ (if factor cancels with denominator and numerator, hole occurs.)

iii) Horizontal asymptotes $\frac{x^2-(a+c)x+c^2}{x^2-(a+b)x+ab} = \frac{x^2}{x^2} = \frac{1}{1} = 1$ (if max. exponent of denominator and numerator are equal, coefficient of two determine horizontal asymptotes.)

iv) Range = {all real numbers except $(y=1)$}

b) $y = \sqrt{2x-1}$

Domain: $2x - 1 \geq 0,\ x \geq \frac{1}{2}$

Range: $y \geq 0$

c) $y = \frac{1}{\sqrt{2x-1}}$ $Domain: 2x - 1 > 0,\ x > \frac{1}{2}$

d) $y = \sqrt{2x^2 - 1}$

Domain $2x^2 - 1 \geq 0 \rightarrow x \leq -\frac{1}{2}, x \geq \frac{1}{\sqrt{2}}$

$\quad\quad y = \sqrt[3]{2x^2 - 1} \rightarrow$ all real numbers

➔ $\quad\quad \sqrt[even]{f(x)} \rightarrow f(x) \geq 0 \rightarrow Domain\ of\ \sqrt[4]{2x - 1} = \ x \geq \frac{1}{2}$

➔ $\quad\quad \sqrt[odd]{f(x)} \rightarrow$ all real numbers $\rightarrow$ Domain of $\sqrt[3]{3x - 1}$ =all real number

22) Statistics

a) (2 , 3, 3, 3, 4, 5, 6, 6, 6, 6, 10, 12)

i) mode =6: a datum with the highest frequency

ii) median =5.5: a middle datum.

$\quad$ (if total number of data is even, then the mean of 2 middle data.)

iii) Range = Max. Value - Min. Value $\Rightarrow 12 - 2 = 10$

iv) mean $= \dfrac{2+3+3+3+4+5+6+6+6+6+10+12}{12} = 5.5$

v) std (standard deviation) $\Rightarrow \sqrt{\dfrac{\sum(each\ value - mean)^2}{n}} =$

$\quad$ n= number of data.

$= \sqrt{\dfrac{(2-5.5)^2+(3-5.5)^2\times3+(4-5.5)^2+(5-5.5)^2+(6-5.5)^2\times4+(10-5.5)^2+(12-5.5)^2}{12}} = 2.84312$

23) Finding inverse function

a) $f(x) = \ln(x^2 - 1) + 1$

inverse function?

$x = \ln(y^2 - 1) + 1$

$x - 1 = \ln(y^2 - 1)$

$y^2 - 1 = e^{x-1}$

$y^2 = e^{x-1} + 1$

$\therefore y = \sqrt{e^{x-1} + 1} \quad\quad$ inverse $f^{-1}(x) = \sqrt{e^{x-1} + 1}$

b) $f(x) = e^{x-1} + 1$

inverse function?

• $x = e^{y-1} + 1$

- $x - 1 = e^{y-1}$

$\ln(x - 1) = \ln e^{(y-1)}$
$\ln(x - 1) = y - 1$
$y = \ln(x - 1) + 1$
$f^{-1}(x) = \ln(x - 1) + 1$

24) Logic

a) $A \rightarrow B$

Above can be expressed as following forms. These are all equivalent.
If A, then B
A implies B
B is necessary for A
A is sufficient for B
B, if A
A only if B

i) Negation (disprove)
$A \rightarrow B'$ or $A \rightarrow B^{\wedge}$ If A, then not B

ii) Contra-positive
$B' \rightarrow A'$
This is the only statement that is logically equivalent to $A \rightarrow B$. This technique is used in INDIRECT PROOF. (This starts with if <u>not B</u> then ends with not A)

iii) Converse: $B \rightarrow A$ (if B, then A) ➔ not always logically equal to $A \rightarrow B$

iv) Inverse: $A' \rightarrow B'$ (If not A, then not B) ➔ not always logically equal to $A \rightarrow B$

25) Vector

Let $\vec{u} = (a, b), and\ \vec{v} = (c, d)$

a) $\vec{u} + \vec{v} = (a + c, b + d)$
$\vec{u} - \vec{v} = (a - c, b - d)$
$\overrightarrow{ku} - \vec{v} = (ka - c, kb - d)$

b) Magnitude of $\vec{u} = \sqrt{a^2 + b^2}$

$$\vec{v} = \sqrt{c^2 + d^2}$$

c) If $\vec{u}$ =initial point (p, q) and terminal point $(r, s), \vec{u} = ((r - p), (s - q))$

d) Unit vector

$$\vec{u} \Longrightarrow (a, b) = \left(\frac{a}{\sqrt{a^2+b^2}}, \frac{b}{\sqrt{a^2+b^2}}\right)$$

26) Geometry Formulas

a) Triangle

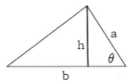

$$h = a \sin\theta$$

Area $= \frac{1}{2}bh = \frac{1}{2}ab \sin\theta$, ($\theta$ is between sides a and b)

b) Trapezoid

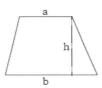

Area $= \frac{h}{2}(a + b)$

c) Circle

Area $= \pi r^2$
Circumference $= 2\pi r$

d) Sector of Circle

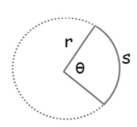

Arc length, $L = r\theta_{radian} = 2\pi r \cdot \frac{\theta^o}{360^o}$

Area $= \frac{1}{2}r^2\theta_{rad} = \pi r^2 \cdot \frac{\theta^o}{360^o}$

d) Right Circular Cone

Volume $= \frac{\pi r^2 h}{3}$

Lateral surface area $= \frac{1}{2}cL = \pi rL$ (where L is lateral surface area)

e) Right Circular Cylinder]

Volume $= \pi r^2 h$
Lateral surface area $= 2\pi r h$

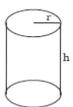

f) Sphere

Volume $= \frac{4}{3}\pi r^3$
Surface area $= 4\pi r^2$

PreCalculus

PreCalculus Notes

This is the main section of this book, and you can learn important parts of PreCalculus by yourself. It is designed to minimize unnecessary problems and study based on the school Textbook, so you can study effectively in preparation for school.

1

Functions

Learning Objective

Know how to identify the Domain and Range of a function and to use function notation to evaluate functions.

Lecture 1.1 Functions

Relation
a set of ordered pairs of quantities that are either related to each other by some or no rule of correspondence.

Function
a relation such that assigns to each input exactly one output.
(Use the vertical line test to test if it is a function)

Domain
the set of numbers that make up all first elements of the ordered pairs
(Input value)

Range
the result set of second elements (Output value)

1. Is the relationship a function?

a.)

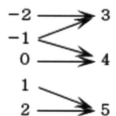

b.) {(2,1), (3,4), (4,1), (5,2), (6.5)}

c.) {(2,0), (3, 1), (2,3)}

d.)

Input Value	0	1	2	1	0
Output Value	-4	-2	0	2	4

e.) $y = 3x - 5$

2. Test the following equations to see if they are functions:

a.) $2x + y^2 = 5$

b.) $x = -3y + 5$

Function Notation: $f(x) = y$

Input	Output	Equation
x	$f(x)$	$f(x) = 8 - 3x$

3. Evaluate the following functions: $g(x) = 8 - 3x$

a.) g(0)=

b.) $g\left(\dfrac{7}{3}\right) =$

c.) g(a+3)=

Piecewise Function: $f(x) = \begin{cases} 2x^2 + 1, x \leq 1 \\ -x^2 + 2, x > 1 \end{cases}$

4. a.) $f(-2)=$

b.) $f(1)=$

c.) $f(2)=$

d.) Is the function continuous?

Find the Domain:

5. a.) $f(x) = 3 - x^2$

b.) $h(x) = \frac{5}{x^2 - 3x}$

c.) $g(x) = \sqrt{x - 4}$

d.) $f(x) = \sqrt{2x} + 1$

Find the difference quotient and simplify:

6. g(x)=2x+1, Find $\frac{g(x+h) - g(x)}{h}$, h ≠ 0

7. $f(x) = 3x^2 - 2, find \ \frac{f(x) - f(7)}{x - 7}, x \neq 7$

Lecture 1.2 Increasing & Decreasing Function, Extrema, Even and Odd

1. Which of the following graphs are graphs of functions?

a.) 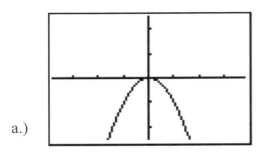   b.)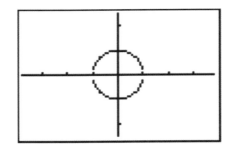

2. Find the domain and the range of $f(x) = \sqrt{x-3}$

Algebraic Solution

Graphical Solution (Graph it)

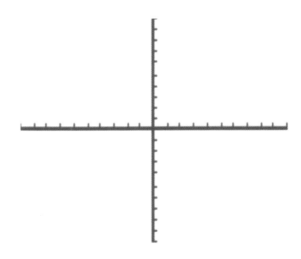

Increasing, Decreasing, and Constant Functions

A function f is increasing on an interval if, for any x_1 and x_2 in the interval, $x_1 < x_2$ implies $f(x_1) < f(x_2)$.

A function f is decreasing on an interval if, for any x_1 and x_2 in the interval, $x_1 < x_2$ implies $f(x_1) > f(x_2)$.

A function f is constant on an interval if, for any x_1 and x_2 in the interval, $f(x_1) = f(x_2)$.

3. Determine the open intervals on which $f(x) = x^3 - 12x$ is increasing or decreasing.

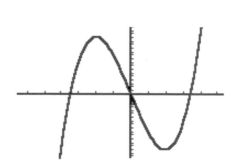

$f(x)$ is increasing on

$f(x)$ is decreasing on

Try another question using your calculator: $f(x) = 2x^3 + 9x^2$

$f(x)$ is increasing on

$f(x)$ is decreasing on

Relative Minimum and Maximum Values

A function value f(a) is called a relative minimum of f if there exists an interval (x_1, x_2) that contains a such that $x_1 < x < x_2$ implies $f(a) \leq f(x)$

A function value f(a) is called a relative maximum of f if there exists an interval (x_1, x_2) that contains a such that $x_1 < x < x_2$ implies $f(a) \geq f(x)$

4. Use your graphing calculator to estimate the relative maximum of $f(x) = -2x^2 - 3x + 1$.

Relative Maximum = _____

5. The profit for a shoe new company can be modeled by $P = .321x^3 - 27.21x^2 + 214x + 135.7$, where P is in thousands of dollars and x is the number of units sold in thousands. What would be the maximum profit for this shoe company?

Maximum profit = _____

The Greatest Integer Function

It is denoted by $[x]$ and is defined by $f(x) = [x]$: the greatest integer less than or equal to x.

6. Graph $f(x) = |x|$

x	$f(x)$

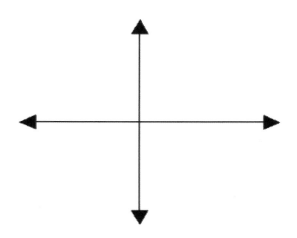

Graphing Piecewise Functions

A piecewise function is a function made up of different parts. More specifically, it's a function defined over two or more intervals rather than with one simple equation over the domain. It may or may not be a continuous function.

7. Sketch the graph of $f(x) = \begin{cases} x^2 + 2, x < 0 \\ 2x + 1, x \geq 0 \end{cases}$

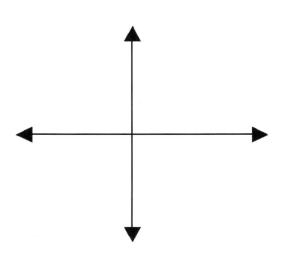

Even and Odd Functions
Even:
- A function f is even if, for each x in the domain of f, $f(-x) = f(x)$.
- The graph of function must be symmetric with respect to the y axis.
- If it is a polynomial function, all exponents of polynomial must be even.

Odd:
- A function f is odd if, for each x in the domain of f, $f(-x) = -f(x)$.
- The graph of function must be symmetric with respect to the origin.
- If it is a polynomial function, all exponents of polynomial must be odd and it should not have a constant.

8. Test the following functions for being odd or even. Use both the algebraic and graphical approach.

Algebraic Solution   Graphical Solution

a.) $f(x) = x^4 - |x|$

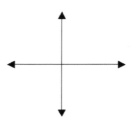

b.) $g(x) = \frac{x}{x^2+1}$

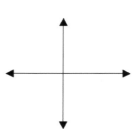

c.) $h(x) = 2x^3 + 4x - 3$

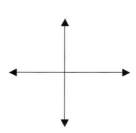

Lecture 1. 3 Shifting, Reflecting and Stretching Graphs

Common functions

constant, identity, absolute value, square root, quadratic, and cubic.

1. Graph the following common functions.

a.) $f(x) = 2$

b.) $f(x) = x$

c.) $f(x) = |x|$

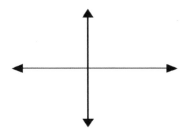
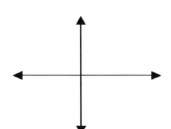
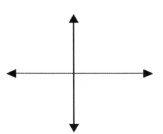

d.) $f(x) = \sqrt{x}$

e.) $f(x) = x^2$

f.) $f(x) = x^3$

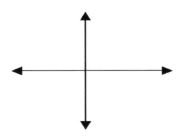
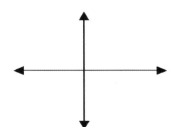
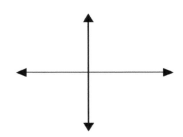

Vertical and Horizontal Shifts

$h(x) = f(x - b)$ produces horizontal shift b units the right.

$h(x) = f(x + b)$ produces horizontal shift b units to the left

$h(x) = f(x) - b$ produces vertical shift b units downward

$h(x) = f(x) + b$ produces vertical shift b units upward

2. $f(x) = x^2$. Describe the shifts in the graph of f generated by the following functions.

a.) $g(x) = x^2 - 3$ _____

b.) $h(x) = (x - 2)^2 + 5$ _____

c.) $j(x) = (x - 6)^2$ _____

Reflections in the Coordinate Axes

1. Reflection in the x-axis: $h(x) = -f(x)$
2. Reflection in the y-axis: $h(x) = f(-x)$
3. Reflection in the x and y-axis: $h(x) = -f(-x)$

3. $f(x) = 2x^3 + 7$. Describe the reflections in the graph of f generated by the following functions:

a.) $g(x) = -2x^3 + 7$

b.) $h(x) = -2x^3 - 7$

4. Sketch the graphs of the four functions by hand on the same rectangular coordinate system.

a.) $f(x) = (x - 2)^2$

b.) $g(x) = (x - 2)^2 + 2$

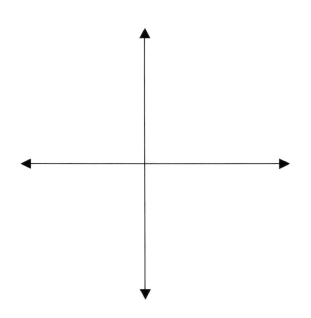

c.) $h(x) = -(x - 2)^2 + 4$

d.) $j(x) = -(-x - 2)^2$

Nonrigid Transformations

Nonrigid transformations are those that cause a distortion – a change in the shape of the original graph.

Nonrigid transformations of $y = f(x)$ come from equations of the form $y = cf(x)$. If $c>1$, then there is a vertical stretch of the graph of $y = f(x)$. If $0<c<1$, then there is a vertical shrink.

5. Graph the following

a.) $f(x) = |x - 1|$ b.) $h(x) = 3|x - 1|$ c.) $g(x) = \frac{1}{3}|x - 1|$

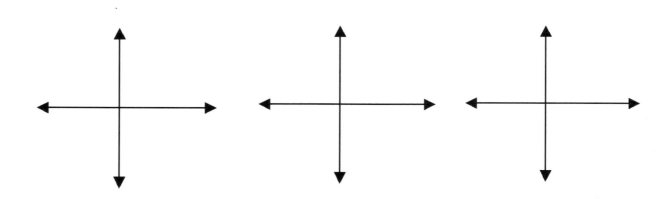

6. Write the equations of the shifts of the following common functions:

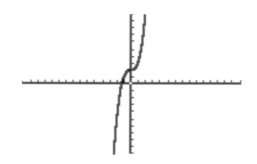 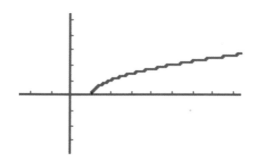

a.) _____   b.)_____

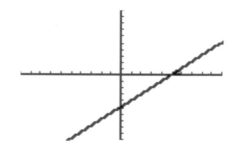 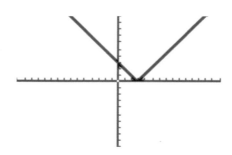

c.)_____   d.)_____

Lecture 1. 4 Combinations of Functions

Sum, Difference, Product, and Quotient of Functions

1. Sum: $(f + g)(x) = f(x) + g(x)$
2. Difference: $(f - g)(x) = f(x) - g(x)$
3. Product: $(fg)(x) = f(x) \cdot g(x)$
4. Quotient: $\left(\frac{f}{g}\right)(x) = \frac{f(x)}{g(x)}, g(x) \neq 0$

1. $f(x) = 3x$ and $g(x) = \sqrt{x - 2}$. Find the following and specify the domain.

a.) $(f+g)(x) =$ domain=_____

b.) $(f-g)(x) =$ domain=_____

c.) $(fg)(x) =$ domain=_____

d.) $\left(\frac{f}{g}\right)(x) =$ domain=_____

e.) $\left(\frac{2f}{g}\right)(x - 1) =$ domain=_____

f.) Evaluate the following using the same f and g.

$(f+g)(3)$ $(f \cdot g)(4)$ $(f-g)(2)$

_____   _____   _____

Definition of Composition of Two functions

The composition of the function f with g is $(f \circ g)(x) = f(g(x))$.
The domain of $f \circ g$ is the set of all x in the domain of g such that $g(x)$ is in the domain of f.

2. Given $f(x) = x^3 + 2$ and $g(x) = \frac{1}{x-1}$ find the following:

a.) $f \circ g$ b.) $g \circ f$

_____   _____

Domain=_____   Domain=_____

3. Find two functions f and g such that $(f \circ g)(x) = h(x)$.

a.) $h(x) = (2 - x)^2$ b.) $h(x) = \sqrt{5 - x}$

$f(x)=$ _____ $f(x)=$ _____

$g(x)=$_____ $g(x)=$_____

4. Use the graphs of f and g to evaluate the functions: (assume 1 interval = 1unit)

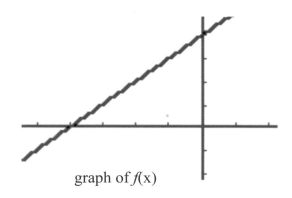

graph of f(x)

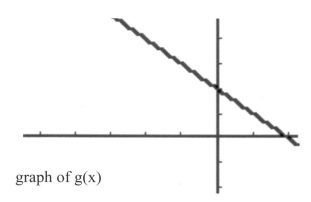

graph of g(x)

a.) $(f+g)(-2)=$_____

b.) $\left(\frac{f}{g}\right)(-1)=$_____

c.) $(fg)(-3)=$_____

c.) $(f \circ g)(0)=$_____

Application Example

5. The number of bacteria in a certain refrigerated food is
$N(T) = 18T^2 - 85T + 400, 0 \leq T \leq 20$, where T is the temperature of the food in degrees Celsius. When the food is removed from refrigeration, the temperature is $T(t) = 3t + 4$, $0 \leq t \leq 4$, where t is the time (in hours). Find the following:

a.) The composite N(T(t)). What does this function represent?

b.) The number of bacteria in the food when t=2 hours.

c.) The time when the bacterial count reaches 2000.

Lecture 1. 5 Inverse Functions

Definition of the Inverse of a Function

Let f and g be two functions such that $f(g(x))=x$ for every x in the domain of g and $g(f(x))=x$ for every x in the domain of f. Under these conditions, the function g is the inverse of the function f. The function g is denoted f^{-1} (read "f-inverse"), So, $f\left(f^{-1}(x)\right) = x$ and $f^{-1}\left(f(x)\right) = x$.

The domain of f must be equal to the range of f^{-1}, and the range of f must be equal to the domain of f^{-1}.

The Graph of an Inverse Function: The graphs of $f(x)$ and $f^{-1}(x)$ are reflections over the line $y=x$.

To find the inverse of a function, replace every x with a y and replace every y with an x and solve the equation for y.

1. Suppose $f = \{(2,1), (3,4), (5,2), (6,7)\}$ find $f^{-1}(x)$.

$f^{-1}(x) =$ _____

2. Find the inverse of $f(x) = \frac{x-1}{4}$.

$f^{-1}(x) =$ _____

3. Verify your result for #2 by showing $f\left(f^{-1}(x)\right) = x$ and $f^{-1}(f(x)) = $ x.

   <u>Verification</u>

4. Graph $f(x) = x^3$ and $g(x) = \sqrt[3]{x}$ and $y = x$ on the same coordinate plane. Are the two functions inverses of each other?

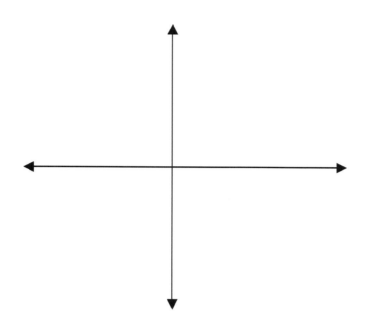

The Existence of an Inverse Function

Not all functions have an inverse over their entire domain.

For example

$f(x) = x^2$ has no inverse over the entire domain. The reason is because $f(1) = f(-1) = 1$, then if you would look at the inverse, but this would not be a function.

To have an inverse, a function must be one-to-one, which means that no two elements in the domain of f correspond to the same element in the range of f. (Horizontal Line Test)

A function f is one-to-one if, for a and b in its domain, $f(a) = f(b)$ implies a=b.

5. Test to see if $f(x) = 2x + 3$ is one-to-one algebraically. If it is one to one, find its inverse.

6. Test to see if $q(x) = 2(x - 3)^2, x \leq 3$ both algebraically and graphically. Find its inverse if you can.

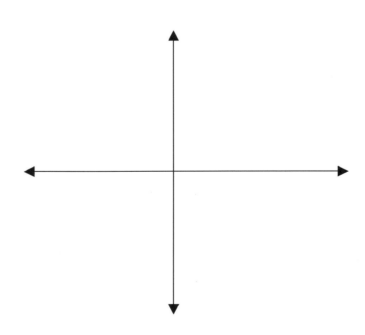

7. Find the inverse of the following functions:

a.) $f(x) = 2\sqrt[3]{x-3}$

Inverse=_____

b.) $g(x) = \dfrac{x-3}{x+1}$

Inverse=_____

2

Polynomial and Rational Function

Learning Objective

Know how to graph and solve Quadratic, Polynomial, and Rational functions and equations.

Lecture 2.1 Quadratic Functions

Constant Function: a polynomial function with degree=0 (ex. $f(x)=5$)

Linear Function: a polynomial function with degree=1 (ex. $f(x)=3x-8$)

Quadratic Function: a polynomial function with degree=2. The graph is always a parabola. (ex. $f(x) = 2x^2 + 3x - 9$)

Definition of a Quadratic Function: Let a, b, and c be real numbers with $a \neq 0$. The function $f(x) = ax^2 + bx + c$ is called a quadratic function.

Standard Form of a Quadratic Function: The quadratic function

$f(x) = a(x - h)^2 + k, a \neq 0$ is said to be in standard form. The graph of *f* is a parabola whose axis is the vertical line x=h and whose vertex is the point (h,k). If a>0, the parabola opens upward, and if a<0, the parabola opens downward.

1. Graph the following function: $f(x) = (x - 1)^2 + 3$ Label the vertex, any intercepts, and the axis of symmetry.

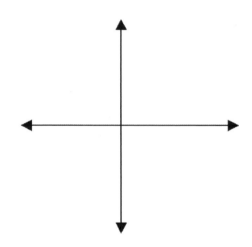

Vertex=_____

x-intercepts (set y=0) | y-intercepts (set x=0)

Note

To graph a parabola, you can start at the vertex and find x and y intercepts. And then connect all 3 points or you can draw a table. Usually vertex, x-intercept, and y-intercepts must be indicated if exists.

2. Sketch the graph of $g(x) = x^2 + 4x + 3$. Label the vertex, any intercepts, and the axis of symmetry.

vertex=_____

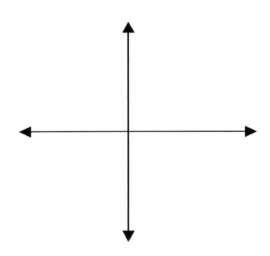

x-intercepts (set y=0) | y-intercepts (set x=0)

3. Write the following functions in vertex form. Identify the vertex of the following quadratic functions. Also state whether the function goes upward or downward.

a.) $h(x) = -3x^2 + 12x - 6$ b.) $f(x) = 4x^2 + 3x - 1$

vertex= _____   vertex= _____

Direction of curve = _____   Direction of curve = _____

4. Graph the following function: $f(x) = -4x^2 + 20x - 8$. Label the vertex, intercepts, and the axis of symmetry.

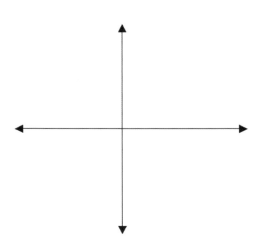

5. Find an equation for the parabola: (vertex= (-2,-9), intercepts=(-5,0) (1,0))

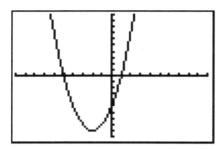

Equation _____

Application

6. The path of a diver is $y = \frac{-3}{8}x^2 + \frac{24}{8}x + 10$ where y is the height (in feet) and x is the horizontal distance (in feet) from the end of the diving board. What is the maximum height of the dive? Verify your answer using a graphing utility.

Lecture 2.2 Polynomial Functions of Higher Degree

Definition of Polynomial Function

Let n be a nonnegative integer and let

$a_n, a_{n-1}, \cdots, a_2, a_1, a_0$ be real numbers with $a_n \neq 0$. The function

$$f(x) = a_n x^n + a_{n-1} x^{n-1} + \cdots + a_2 x^2 + a_1 x + a_0$$

is called a polynomial function of x with degree n.

The graph of a polynomial function is continuous. (meaning that it has no breaks, holes, or gaps)

1. Find the degree of the following polynomial functions:

a.) $f(x) = 3x^4 + 2x - 6$

b.) $f(x) = 2x^3 - 3x + 5x^6$

Degree=_____

Degree=_____

2. Label the graphs of $y = x^2, y = x^4,$ and $y = x^8$.

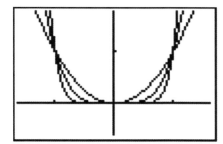

In the graphs of $f(x) = x^n$ where n was even, as n increases the graph flattens at the origin.

3. Label the graphs of $y = x^3, y = x^5,$ and $y = x^7$.

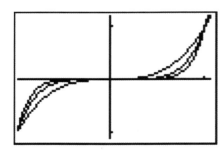

In the graphs of $f(x) = x^n$ where n was odd, as n increases the graph flattens at the origin.

4. Sketch the graph of $f(x) = -(x+3)^4 - 2$

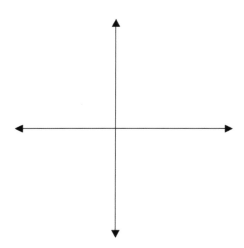

5. Sketch the graph of $f(x) = (x-2)^5 + 1$

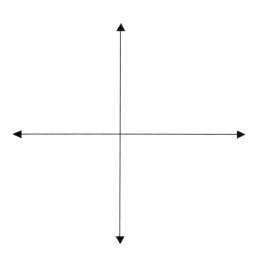

The Leading Coefficient Test

Determines whether the graph of a polynomial eventually rises or falls by examining the function's degree (even or odd) and by its leading coefficient.

$f(x) = a_n x^n + \cdots$	$a_n > 0$	$a_n < 0$
n even	↖ ↗	↙ ↘
n odd	↙ ↗	↖ ↘

6. Describe the right-hand and left-hand behavior of the graph of each function.

a.) $f(x) = -2x^4 + 8x^3 - 14x - 2$ _____

b.) $g(x) = 3x^5 - x^3 - 12x^2 + 5$ _____

Zeros of Polynomial Functions

If f is a polynomial function and a is a real number, the following statements are equivalent.

1. $x=a$ is a zero of the function f
2. $x=a$ is a solution of the polynomial equation $f(x)=0$
3. $(x-a)$ is a factor of the polynomial $f(x)$
4. $(a, 0)$ is an x-intercept of the graph of f.

* The graph of f with degree n has at most n real zeros.
* The graph of f with degree n has at most n-1 extrema (relative minimums or maximums)

7. Find the x-intercepts of the graph of $f(x) = x^3 - 2x^2 - 4x + 8$ and graph.

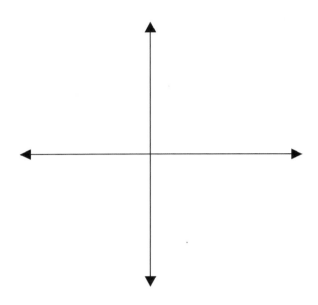

8. Sketch the graph of $f(x) = 2x^3 - 4x^2$.

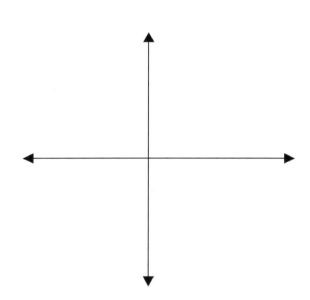

9. Find a polynomial function with the following zeros. (assume leading Coefficient =1)

a.) -2, 1, 4

The Intermediate Value Theorem

The Intermediate Value Theorem concerns the existence of real zeros of polynomial functions. The theorem states that if $(a, f(a))$ and $(b, f(b))$ are two points on the graph of a polynomial function such that $f(a) \neq f(b)$, then for any number d between $f(a)$ and $f(b)$ there must be a number c between a and b such that $f(c) = d$.

---

10. Use the Intermediate Value theorem and a graphing utility to find intervals of length 1 in which the polynomial function is guaranteed to have a zero. Then approximate the zeros of the function to one-tenth.

a.) $h(x) = 28x^3 - 11x^2 + 15x - 28$

11. Match the equations with the graphs below:

a.) $y = -x^3$ b.) $y = x^2$ c.) $y = x^4$ d.) $y = -x^4$

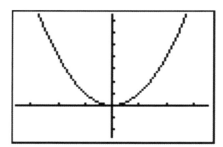

1._____

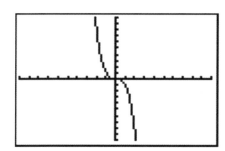

2._____

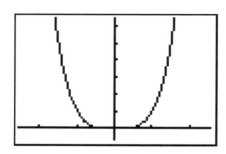

3._____

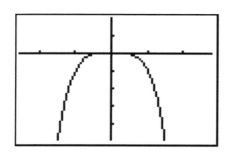

4._____

Lecture 2.3 Real Zeros of Polynomial Functions

Long Division of Polynomials

An algorithm for dividing a polynomial by another polynomial of the same or lower degree

1. Divide $3x^3 - 4x^2 + x - 7$ by $x - 2$

$$x-2 \overline{)3x^3 - 4x^2 + x - 7}$$

Division Algorithm

If $f(x)$ and $d(x)$ are polynomials such that $d(x) \neq 0$, and the degree of $d(x)$ is less than or equal to the degree of f(x), there exist unique polynomials $q(x)$ and $r(x)$ such that

$$f(x) = d(x)q(x) + r(x)$$

f(x) is the dividend
d(x) is the divisor
q(x) is the quotient
r(x) is the remainder

where r(x)=0 or the degree of r(x) is less than the degree of d(x). If the remainder r(x) is zero, d(x) divides evenly into $f(x)$.

2. Divide $4x^3 - x^2 + 5x - 3$ by $x^2 - 2x + 3$

Synthetic Division

Short form of long division of divisors of the form (x-k)

3. Use synthetic division to divide $3x^4 - 2x^2 - x + 1$ by $x - 3$

The Remainder Theorem

If a polynomial $f(x)$ is divided by $x - k$, the remainder is $r = f(k)$

4. Use the remainder theorem to evaluate $f(x) = 3x^2 - 8x - 2$ when $x = 3$

The Factor Theorem

A polynomial $f(x)$ has a factor $(x - k)$ if and only if $f(k)=0$.

5. Determine whether or not x –2 is a factor of $f(x) = x^4 - 16$ using the factor theorem.

The Rational Zero Test

If the polynomial $f(x) = a_n x^n + a_{n-1} x^{n-1} + \cdots + a_2 x^2 + a_1 x + a_0$ has integer coefficients, every rational zero of f has the form Rational zero $= \frac{p}{q}$, where p and q have no common factors other than 1, p is a factor of the constant term a_0, and q is a factor of the leading coefficient a_n.

6. Solve $x^3 - 2x^2 - 5x + 6 = 0$.

P: q: $\frac{p}{q} =$

Use synthetic division to find the solutions from the $\frac{p}{q}$ set.

Solutions:_____

7. Find all the real zeros of $2x^3 - 3x^2 - 9x + 10 = 0$.

Zeros_____

8. Find all the zeros: $f(x) = x^4 - 6x^3 + 6x^2 + 10x - 3$

Zeros_____

9. Find all the zeros: $f(x) = 2x^4 - x^3 - 8x^2 - x - 10$

Zeros_____

Lecture 2.4 Complex Numbers

Some quadratic equations have no real solutions, like $x^2 + 1 = 0$. To solve this problem, mathematicians created an expanded system of numbers using the imaginary unit i.

$i = \sqrt{-1}$, where $i^2 = -1$.

Definition of a Complex Number

If a and b are real numbers, the number $a + bi$ is a complex number, and it is said to be written in standard form.
If b=0, the number $a + bi = a$ is a real number.
If $b \neq 0$, the number $a + bi$ is called an imaginary number.
A number of the form bi, where $b \neq 0$, is called a pure imaginary number.

$a + bi = c + di$ if and only if a=c and b=d.

Sum: $(a + bi) + (c + di) = (a + c) + (b + d)i$
Difference: $(a + bi) - (c + di) = (a - c) + (b - d)i$

1. Simplify the following complex numbers by adding or subtracting:

a.) $(5 - 2i) + (4 + 3i)$

b.) $(5 - i) - (2 - 4i)$

2. Multiply the following complex numbers:

a.) $3(15 - 7i) =$

b.) $(2 - i)(4 + 3i) =$

c.) $(2 + 3i)^2 =$

d.) $(3 + 5i)(1 - 5i) =$

Complex Conjugates

$a + bi$ and $a - bi$ are called complex conjugates. $(a + bi)(a - bi) = a^2 + b^2$

3. Divide the following complex numbers:

a.) $\dfrac{3}{2 - i} =$

b.) $\dfrac{3 - i}{2 + 3i} =$

Plotting Complex Numbers

4. Plot each of the complex numbers in the complex plane.

a.) $4 + 2i$ b.) $5i$ c.) -3 d.) $-2 - 4i$ e.) $i - 5$

Real Axis

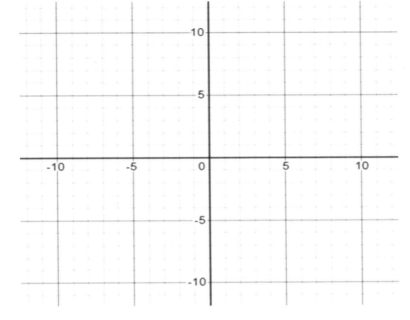

Imaginary Axis

Powers of i. (4 possibilities)

$i^1 = i$

$i^2 = -1$

$i^3 = -i$

$i^4 = 1$

$i^5 = i$ ➜ it repeats (same as i^1)

$i^6 = -1$➜ it repeats (same as i^2)

5. Express each of the following powers of i as $i, -i, 1$ *or* -1

a.) $i^{16} =$

b.) i^{25}

c.) i^{102}

d.) i^{-19}

6. Write the following in standard form:

a.) $5 + \sqrt{-28} =$

b.) $\sqrt{-100} + 4i - 5 =$

7. Solve for a and b.

a.) $(a + 2) + (b - 3)i = 6 + 10i$

a=_____   b=_____

Lecture 2.5 The Fundamental Theorem of Algebra

In the complex number system, every <u>n</u>th-degree polynomial function has precisely <u>n</u> zeros. This result is derived from the Fundamental Theorem of Algebra.

The Fundamental Theorem of Algebra

If $f(x)$ is a polynomial of degree n, where $n>0$, f has at least one zero in the complex number system.

Linear Factorization Theorem

If $f(x)$ is a polynomial of degree n where $n>0$, f has precisely n linear factors

$$f(x) = a_n(x - c_1)(x - c_2) \cdots (x - c_n)$$

where $c_1, c_2, \cdots, c_n$ are complex numbers.

1. Solve $x^3 + 5x - 6 = 0$

Solutions_____

2. Find all real zeros of $f(x) = x^4 - 5x^3 + x - 5$

Zeros_____

Complex Zeros Occur in Conjugate Pairs

Let f(x) be a polynomial function that has real coefficients. If a + bi, where $b \neq 0$, is a zero of the function, the conjugate a − bi is also a zero of the function.

3. Find a fourth-degree polynomial function with real coefficients that has 0, 2, and i as zeros. (assume leading Coefficient =1)

4. Factor $f(x) = x^4 - 12x^2 - 13$

a.) as the product of factors that are irreducible over the rationals.

b.) as the product of factors that are irreducible over the reals.

c.) completely.

d.) List the solutions:

5. Find all zeros of $f(x) = x^4 + 4x^3 + 7x^2 + 16x + 12$, given that $2i$ is a zero.

6. Given $f(x) = x^5 - 7x^4 + 13x^3 - 31x^2 + 36x - 12$.

Calculator question.
a.) Find all zeros of the function b.) Write the polynomial as a product of linear factors, c.) use your factorization to determine the x-intercepts of the graph of the function, and d.) use a graphing utility to verify that the real zeros are the only x-intercepts.

Lecture 2.6 Rational Functions and Asymptotes

A rational function can be written in the form $f(x) = \frac{N(x)}{D(x)}$, where N(x) and D(x) are polynomials and D(x) is not the zero polynomial.

Reminder: Domain: (denominator of a fraction cannot equal zero)

Vertical Asymptotes

Restrictions on Domain (except when the factor is in the numerator, then it is a hole)

Horizontal Asymptotes

The value as f(x) as x approaches negative infinity and positive infinity. (What happens when x gets very large or very small? There could be two different horizontal asymptotes)

1. Find the domain of $f(x) = \frac{1}{x+2}$ Then sketch a graph.

Vertical Asymptote:_____

Horizontal Asymptote:_____

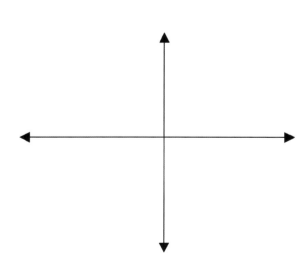

Method to identify the horizontal asymptotes

$f(x) = \frac{N(x)}{D(x)}$. Let n=the degree of N(x) and let m=the degree of D(x).

a.) If n<m, then y=0.

b.) If n=m, then $y = \frac{(leading\ coefficient\ of\ N(x))}{(leading\ coefficient\ of\ M(x))}$.

c.) If n>m, then the graph has no horizontal asymptotes. To find the slant asymptote, use long division.

2. Given: $f(x) = \frac{2x}{x+1}$ Determine the horizontal and vertical asymptotes and then graph.

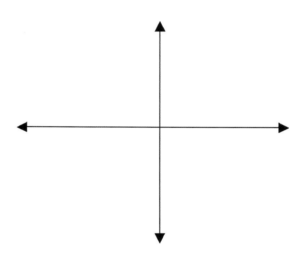

3. Given: $f(x) = \frac{3x}{x^2-9}$ Determine the horizontal and vertical asymptotes and then graph.

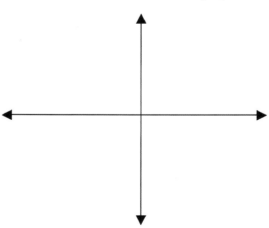

4. Given: $f(x) = \dfrac{x^2}{x+2}$ Determine the horizontal and vertical asymptotes and then graph.

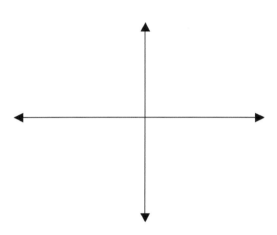

5. Given: $f(x) = \dfrac{x}{|x-2|}$ Determine the horizontal and vertical asymptotes and then graph.

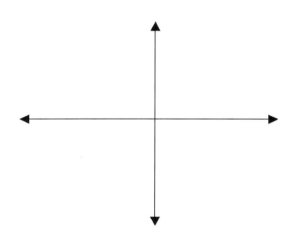

6. Given $f(x) = \dfrac{x-3}{x^2-7x+10}$ Determine the horizontal and vertical asymptotes and then graph.

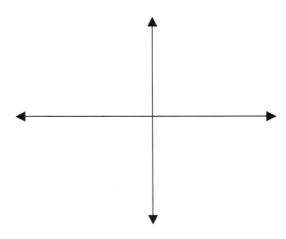

7. The Environmental Protection Organization has determined that if 800 deer are introduced to a preserve, the population at any time t (in months) is given by $N = \dfrac{800+350t}{1+0.2t}$ What is the carrying capacity of the preserve?

Lecture 2.7 Graphs of Rational Functions

When Sketching the graph of a rational function

1.) Find and plot the x and y intercepts (if any) set x and y equal to 0.
2.) Find the vertical asymptotes. (set denominator equal to 0.)
3.) Find the horizontal asymptotes.
4.) Plot at least one point between and one point beyond each x-intercept and vertical asymptote.

Symmetry Tests

a.) if $f(-x)=f(x)$, then symmetrical with respect to y-axis.
b.) if $-f(x)=f(-x)$, then it is symmetrical with the origin.

1. Sketch the graph of $f(x) = \frac{1}{x-3}$ Identify the following:

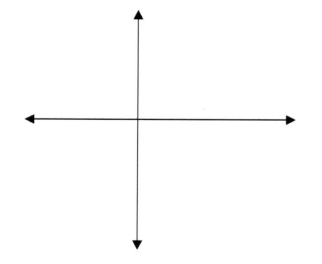

x-intercepts:_____

y-intercepts:_____

vertical asymptotes:_____

horizontal asymptotes:_____

Symmetry:_____

Additional checking Points

X					
$f(x)$					

2. Sketch the graph of $g(x) = \dfrac{x^2}{x^2-4}$

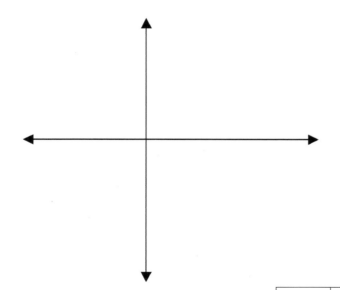

x-intercepts:_____

y-intercepts:_____

vertical asymptotes:_____

horizontal asymptotes:_____

Symmetry:_____

Additional Points

x					
g(x)					

3. Sketch the graph of $g(x) = \dfrac{x^2+2}{x-1}$

x-intercepts:_____

y-intercepts:_____

vertical asymptotes:_____

horizontal asymptotes:_____

slant asymptotes:_____

Symmetry:_____

Additional Points

x					
g(x)					

4. Sketch the graph of $f(x) = \frac{x^2 - 5x + 4}{x - 3}$.

x-intercepts:_____

y-intercepts:_____

vertical asymptotes:_____

horizontal asymptotes:_____

slant asymptotes:_____

Symmetry:_____

Additional Points

x					
$f(x)$					

$f(x) = \frac{x^2 - 5x + 4}{x - 3}$

3

Exponent and Log

Learning Objective

Know how to graph Exponent and Logarithmic functions and solve the equations.

Lecture 3.1 Exponential Functions and Their Graphs

Definition of Exponential Function

The exponential function f with base a is denoted by $f(x) = a^x$ where $a > 0, a \neq 1$, and x is any real number.

1. Graph the following exponential function: $f(x) = 2^x$

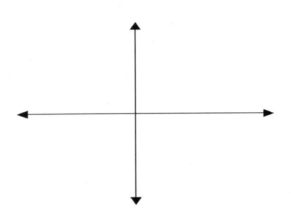

Domain:_____

Range:_____

y-intercept:_____

horizontal asymptote:_____

additional points: _____

2. Graph the following exponential function: $f(x) = 2^{-x}$

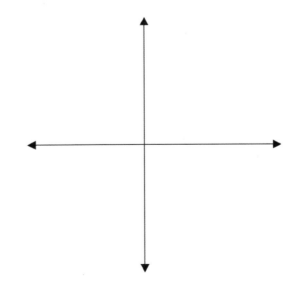

Domain:_____

Range:_____

y-intercept:_____

horizontal asymptote:_____

additional points: _____

3. Graph each of the following on the same coordinate axis:

a.) $g(x) = 3^x$

b.) $f(x) = 3^{x-2}$

c.) $h(x) = 3^{x-2} + 4$

*Use the transformation rules

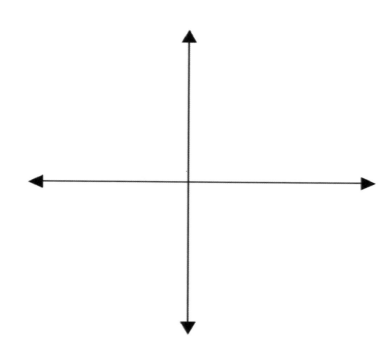

The Natural Base e

An irrational number (Euler's number) $e \approx 2.7182818\ldots$

4. Use a calculator to complete the following table:

x	1	10	100	1000	10000	100000
$\left(1 + \dfrac{1}{x}\right)^{x}$						

Note that $\left(1 + \dfrac{1}{x}\right)^{x}$ as $x \to \infty$.

5. Graph $f(x) = e^{x}$

Domain:_____

Range:_____

Intercepts: _____

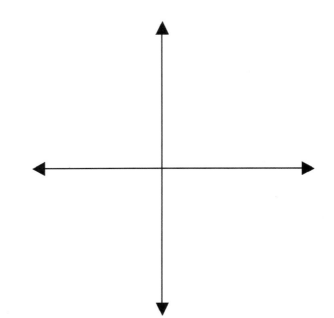

Formulas for Compound Interest

After t years, the balance $A(t)$ in an account with principal A_0 and annual interest rate r (expressed as a decimal) is given by the following formulas.

1. For $\underline{n}$ compoundings per year : $A(t) = A_0\left(1 + \dfrac{r}{n}\right)^{nt}$

2. For continuous compounding: $A(t) = A_0\, e^{rt}$

$A(t)$ is the amount in the account after t years.
A_0 is the principal amount (the original amount)
r is the annual interest rate (as a decimal)
n is the number of pay periods per year. (number of times interest is calculated)

6. An investment of \$4000 is made into an account that pays 2% annual interest for 10 years. Find the amount in the account if the interest is compounded:

a.) annually. n=1   b.) quarterly. n=4   c.) monthly. n=12   d.) daily, n=365

1	4	12	365
$	$	$	$

e.) What would the amount equal if interest is compounded continuously? _____

7. The population of a city increases according to the model $P(t) = 29{,}000e^{0.0147t}$, where t=0 corresponds to 1980. Use this model to predict the population in 2008.

Predicted population in 2008=_____

8. Let Q(t) (in grams) represent the mass of a quantity of carbon 14, which has a half-life of 5710 years. The quantity present after t years is $Q(t) = 50\left(\frac{1}{2}\right)^{\frac{t}{5710}}$.

a.) Determine the initial quantity (when t=0)
b.) Determine the quantity present after 3000 years.
c.) Sketch the graph of the function over the interval t=0 to t=10,000

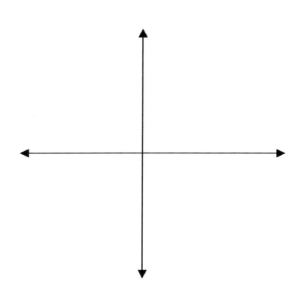

Lecture 3.2 Logarithmic Functions and Their Graphs

Since the exponential function of the form $f(x) = a^x$ is one-to-one (meaning it passes the horizontal line test), then its inverse must also be a function. This inverse function is called the logarithmic function with base a.

Definition of Logarithmic Function

For $x > 0$ and $0 < a \neq 1$,

$$y = \log_a x \text{ if and only if } x = a^y.$$

The function $f(x) = \log_a x$ is called the logarithmic function with base a.

1. Evaluate each of the following:

a.) $\log_2 4 =$

b.) $\log_2 0.125 =$

c.) $\log_3 27 =$

d.) $\log_{10} 1000 =$

(Note: we typically write $\log_{10} 1000 =$ as $\log 1000 =$ since the calculator use log base 10, it is called the common logarithm.)

e.) $\log_{10}(-1.2) =$

f.) $\log_7(-43) =$

Properties of Logarithms

1. $\log_a 1 = 0$ because $a^0 = 1$
2. $\log_a a = 1$ because $a^1 = a$.
3. $\log_a a^x = x$ and $a^{\log_a x} = x$.
4. If $\log_a x = \log_a y$, then x=y

2. Solve the following equations:

a.) $log_5 x = log_5 12$

b.) $log_4 1 = x$

Graphs of Exponential Functions

3. Sketch the graph of the following on the same coordinate axis.

a.) $y = log_{10} x$

b.) $y = log_{10}(x + 3)$

c.) $y = log_{10}(x + 3) - 2$

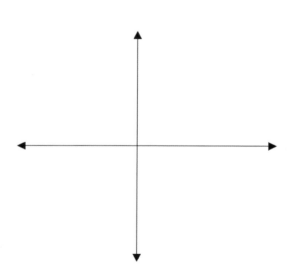

The Natural Logarithmic Function

The function defined by $f(x) = \log_e x = \ln x$, $x > 0$ is called the natural logarithmic function.

Properties of Natural Logarithms

1. $\ln 1 = 0$ because $e^0 = 1$.

2. $\ln e = 1$ because $e^1 = e$.

3. $\ln e^x = x$ and $e^{\ln x} = x$.

4. If $\ln x = \ln y$, then $x = y$.

4. Evaluate the following:

a.) $\ln e^3 =$

b.) $e^{\ln 7} =$

c.) $\ln \frac{1}{e^4} =$

5. Use your calculator to evaluate each expression:

a.) ln 5.3 =   b.) ln 1.4=   c.) ln (-3.6)=

6. Find the domain of the following functions:

a.) $f(x) = \ln(x + 5)$ b.) $f(x) = 2 \ln|x|$

Application Problem.

7. The model $t = 12.542 \ln\left(\frac{x}{x-1000}\right), x > 1000$ approximates the length of a home mortgage of $149,000 at 7.8% in terms of the monthly payment. In the model, t is the length of the mortgage in years and x is the monthly payment in dollars. Find the length of the home mortgage of $149,000 at 7.8% if the monthly payment is $2000 and the total interest charged over the life of the loan.

Lecture 3.3 Properties of Logarithms

Change of Base

- most calculators can only evaluate common logarithms (base 10) and natural logarithms (base e)

To evaluate a logarithm of a different base, you can use the change-of-base formula

Change-of-Base Formula

Let a, b, and x be positive real numbers such that $a \neq 1$ and $b \neq 1$. Then $log_a x$ can be converted to a different base using any of the following formulas:

Base b	Base 10	Base e
$log_a x = \dfrac{log_b x}{log_b a}$	$log_a x = \dfrac{log_{10} x}{log_{10} a}$	$log_a x = \dfrac{\ln x}{\ln a}$

1. Evaluate the following:

a.) $log_4 32 =$

b.) $log_3 51 =$

Properties of Logarithms

Let a be a positive number such that $a \neq 1$, and let n be a real number. If u and v are positive real numbers, the following properties are true.

1. $\log_a(uv) = \log_a u + \log_a v$

2. $\log_a \frac{u}{v} = \log_a u - \log_a v$

3. $\log_a u^n = n\log_a u$

1. $\ln(uv) = \ln u + \ln v$

2. $\ln \frac{u}{v} = \ln u - \ln v$

3. $\ln u^n = n\ln u$

2. Expand the logarithmic expressions:

a.) $\log(5x^2 y^5) =$

b.) $\ln \dfrac{\sqrt{x+3}}{y^4} =$

c.) $\ln \dfrac{2x\sqrt{x-2}}{y^{4/3}}$

3. Condense the logarithmic expression:

a.) $3 \log x - 5 \log y + \frac{1}{2} \log(3z) =$

b.) $\frac{2}{3}(3 \ln x - 5 \ln y + \ln(z - 5)) =$

c.) $\frac{1}{5}(4 \ln x + 3 \ln 5 \cdot \ln(z + 1)) =$

4. Rewrite the logarithm as a multiple of a) common logarithm and b.) a natural logarithm

$log_3 x$ a.)   b.)

5. Solve the following problems:

a.) $(log_2 x)^2 = 16$ b.) $log_2 x^2 = 16$

c.) $log_5(x - 3) + log_5(x + 3) = 4$ d.) $3^{2x} = 243$

e.) $8^{(2x-1)} = 4^{(x-3)}$

f.) $3^{(2x-4)} = 12^{(3x+1)}$

g.) $7^{(3x-2)} = \ln 2 \cdot 7^{(2x+5)}$

Lecture 3.4 Solving Exponential and Logarithmic Equations

Strategies for Solving Exponential and Logarithmic Equations

1. Rewrite the given equation in a form to use the One-to-One Properties of exponential or logarithmic functions.

2. Rewrite an exponential equation in logarithmic form and apply the Inverse Property of logarithmic functions.

3. Rewrite a logarithmic equation in exponential form and apply the Inverse Property of exponential functions.

Review of one-to-one and inverse properties:

One-to-One Properties

$a^x = a^y$ if and only if x=y.

$log_a x = log_a y$ if and only if x=y.

Inverse Properties

1.) $log_a a^x = x$

2.) $a^{log_a x} = x$

1. Solve each equation and round your answer to three decimal places.

a.) $4e^{2x} = 16$

x=_____

b.) $5e^{x+2} - 8 = 14$

$x=$ _____

c.) $2(3^x - 1) = 10$

$x=$ _____

d.) $e^{2x} - e^x - 20 = 0$

$x=$ _____

Solving Logarithmic Equations

Two possible ways

a.) Isolate the logarithmic expression and then write the equation in equivalent exponential form.

b.) Get a single logarithmic expression on each side of the equation, with the same base, then use the one-to-one property.

2. Solve the following logarithmic equations and round your answer to three decimal places.

a.) $3\log x = 5$

X=_____

b.) $\ln\sqrt{x+3} = \ln x$ (check for extraneous solutions)

X=_____

c.) $\log x - \log(x-2) = 1$

X=_____

Approximating Solutions

Some equations are beyond our algebraic skills and only approximate solutions can be obtained.

3. Approximate the solution of $2\ln(x-1) = -x^2 + 9$. Graph both sides of the equation and look for points of intersection. (Use the Calculator's intersect option on the graphing calculator)

Make a sketch of the graph and write your answer to 5 decimal places.

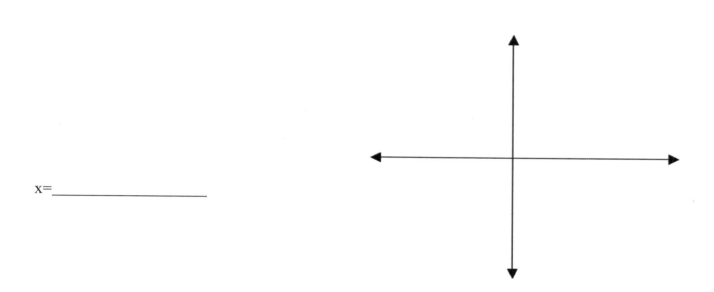

x=_____

Application

4. How long would it take for an investment to double if the interest was compounded continuously at 3.5%? $(A(t) = Ae^{rt})$

5. You have $80,000 to invest. You need to have $500,000 to retire in thirty years. At what continuously compounded interest rate would you need to invest at to reach your goal?

$(A(t) = Ae^{rt})$

6. If $500 is invested at 6%, compounded continuously, how long (to the nearest year) will it take for the money to triple?

Lecture 3.5 Exponential and Logarithmic Models

Five different mathematical models

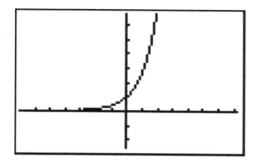

1. Exponential growth model: $y = ae^{bx}, b > 0$

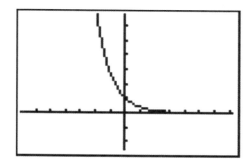

2. Exponential decay model: $y = ae^{bx}, b < 0$

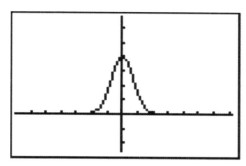

3. Gaussian model: $y = ae^{\frac{-(x-b)^2}{c}}$

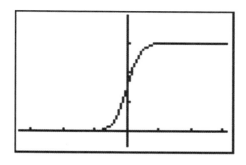

4. Logistic growth model: $y = \frac{a}{1+be^{-rx}}$

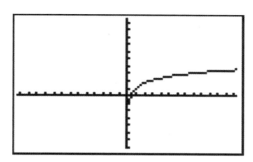

5. Logarithmic models: $y = a + b \cdot \ln x, \quad y = a + b \cdot log_{10} x$

1. The population of a large city can be modeled by $y = 1.95e^{0.0226x}$, in millions, where x=0 corresponds to 1995. In what year is the population of this city expected to reach 2.8 million?

2. Two hours after bacteria were introduced to a culture the population was 100. Five hours after that the population was 400. What will the population be 24 hours after the start of the experiment? (Assume, this is an exponential growth model)

3. The radioactive isotope ^{226}Ra has a half-life of about 1619 years. If the original amount introduced was 40 grams, how much would remain after 5,000 years? (half-life is the amount of time required for one-half of an original amount to decay.)

4. Use a graphing utility to fit a logarithmic model to the following data:

x	2	3	4	5	10	15	20
y	3.16	4.38	5.24	5.91	8.00	9.22	10.09

Use the regression capabilities of the graphing calculator to find the logarithmic model.

5. The table below shows the profit for a company (in millions of dollars) for the years 1990 (x=0) to 1998. Use a graphing calculator to fit an exponential model to the data, then use the model to predict the company's profit in 2005.

T	0	1	2	3	4	5	6	7	8
P	5.80	6.35	6.94	7.60	8.31	9.10	9.95	10.89	11.92

Use the regression capabilities of the graphing calculator to find the exponential model.

Exponential Model_____

Profit in 2005 = _____

4

Trigonometry

Learning Objective

Know how to measure an angle, draw a Unit Circle, and calculate Trig. values without a calculator.
Know how to graph all Trig. Functions.

Lecture 4.1 Measure of an Angle: Angle, DMS, Radian vs Degree

Definition

Initial side: The starting position of the angle. (usually from + x-Axis)

Terminal side: The ending position of the angle.

Standard Position: An angle is in standard position if its vertex is at the origin and the initial side coincides with the positive x-axis.

Positive Angles: Generated by counterclockwise rotation.

Negative Angles: Generated by clockwise rotation.

Measure of an Angle:

Radians: One radian is the measure of a central angle θ that intercepts an arc s equal in length to the radius r. $(\theta = \frac{s}{r})$

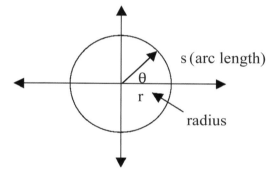

1 radian $= 57.3^0$

$\frac{\pi}{2}$ radian $= 90^0$

π radians $= 180^0$

2π radians $= 360^0$

Degrees: 1° equals $\frac{1}{360}$ of a revolution

Degrees

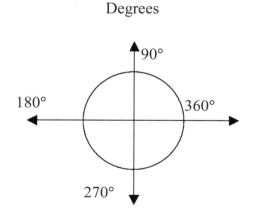

Radians

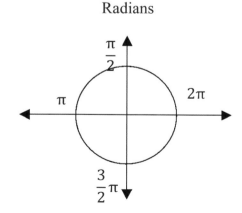

Coterminal Angles

Two angles that have the same initial and terminal sides.

To find a coterminal angle: Either Add 2π or $360°$ or Subtract 2π or $360°$ from the angle that is given

1. Find a coterminal angles for each of the following:

a.) $\frac{11\pi}{6}$

b.) $\frac{-\pi}{4}$

c.) $180°$

d.) $-150°$

e.) $540°$

2. Determine the quadrant where the terminal side of each angle lies:

a.) $\frac{916\pi}{6}$

b.) $\frac{-10\pi}{3}$

c.) $\frac{97\pi}{4}$

d.) $\frac{-17\pi}{3}$

Complementary Angles

Two angles are complementary if the sum of their angles *is* $\frac{\pi}{2}$ or 90°

Supplementary Angles

Two angles are supplementary if the sum of their angles *is* $\frac{\pi}{2}$ or 90°

3. Find the complement and supplement of each of these:

a.) $\frac{4\pi}{7}$

b.) $\frac{1\pi}{8}$

c.) $\frac{1\pi}{10}$

Degrees, Minutes, and Seconds

$1° = 60'$ (60 minutes) and $1' = 60''$ (1 minute=60 seconds)

4. Convert to decimal degrees:
a.) 125° 12'   b.) -100° 34'

5. Convert to degrees and minutes:
a.) 52.6°   b.) 284.18°

To Convert Degrees to Radians:	$\theta_r = \theta° \times \dfrac{\pi}{180°}$
To Convert Radians to Degrees:	$\theta° = \theta_r \times \dfrac{180°}{\pi}$

6. Convert from degrees to radians:

a.) 225°   b.) -120°

7. Convert from radians to degrees:

$a.$) $\dfrac{15\pi}{6}$ b.) $\dfrac{-5\pi}{3}$

Arc Length (s):

$$s = r \times \theta_r \quad \text{where } \theta \text{ is in radians}$$

Area of sector of a circle:

$$A = \frac{1}{2}r^2\theta_r \quad \text{where } \theta \text{ is in radians}$$

8. Find the area of the sector and the length of the arc subtended by a central angle of $\frac{2\pi}{3}$ radians in a circle whose radius is 6 inches.

9. Find the length of the arc on a circle of radius 30 centimeters intercepted by a central angle of 60°.

10. Find the radian measure of the central angle of a circle of radius 22 feet that intercepts and arc of length 10 feet.

11. If a sector of a circle has an arc length of 3π inches and an area of 9π square inches, what is the length of the radius of the circle?

Lecture 4.2 Unit Circle and Special Angle

To develop the unit circle:

#1   Label each quadrant inside the circle with I, II, III, and IV

#2   Label the points on the x and y axis with their ordered pairs.

#3   Write the ordered pairs that are in each quadrant, i.e. (+, +)

#4   Use the 45 – 45 – 90 triangle … find the appropriate ordered pairs

#5   Label angles in degrees

#6   Use the 30 – 60 – 90 triangle … find the appropriate ordered pairs

#7   Double check all your signs on your ordered pairs.

#8   Label radians on each tic mark

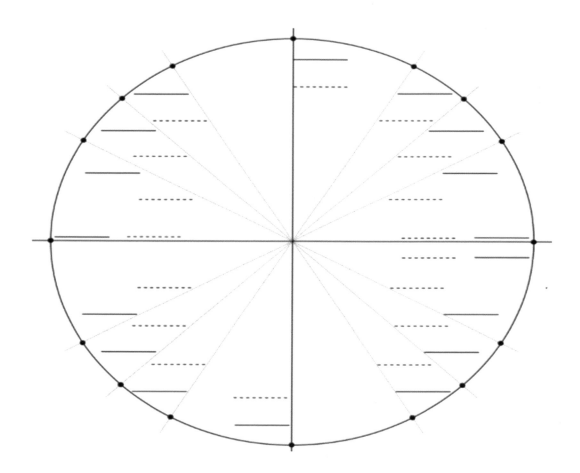

Let t be a real number and let (x,y) be the point on the unit circle corresponding to t.

$\cos t = x$ $\qquad \sin t = y$

$\tan t = \dfrac{y}{x}, x \neq 0$ $\qquad \csc t = \dfrac{1}{y}, y \neq 0$

$\sec t = \dfrac{1}{x}, x \neq 0$ $\qquad \cot t = \dfrac{x}{y}, y \neq 0$

1, Evaluate the six trigonometric functions at each real number:

	sin t	cos t	tan t	cot t	sec t	csc t
	y	x	$\dfrac{y}{x}$	$\dfrac{x}{y}$	$\dfrac{1}{x}$	$\dfrac{1}{y}$
$\dfrac{7\pi}{4}$						
$\dfrac{5\pi}{6}$						
$\dfrac{-2\pi}{3}$						
$\dfrac{13\pi}{6}$						
$\dfrac{-7\pi}{2}$						

A function is called periodic if it repeats itself over and over again at regular intervals. Mathematically, a function f is periodic if there exists a positive real number c such that $f(t+c) = f(c)$ for all t in the domain of f. The smallest number c for which f if periodic is called the period of f.

Recall: f is an even function if $f(-x) = f(x)$, f is an odd function if $f(-x) = -f(x)$

The cosine and secant functions are even.

$\cos(-\theta) = \cos(\theta)$ $\sec(-\theta) = \sec(\theta)$

The sine, cosecant, tangent, and cotangent are odd.

$\sin(-\theta) = -\sin(\theta)$ $\csc(-\theta) = -\csc(\theta)$

$\tan(-\theta) = -\tan(\theta)$ $\cot(-\theta) = -\cot(\theta)$

2. Use the Value of the trig function to evaluate the indicated functions.

a.) $\sin(-\theta) = 3/5$ Find   (i) $\sin\theta$ (ii) $\csc\theta$

b.) $\cos(-\theta) = \dfrac{2}{7}$ Find   (i) $\sin\theta$ (ii) $\sec\theta$

3. Domain of Sine and Cosine: 4. Range of Sine and Cosine:

Lecture 4.3 Trigonometry Definition

Remember:

*Sin(angle)= Ratio of right triangle.

$\sin\theta=\dfrac{opp}{hyp}$ $\csc\theta=\dfrac{hyp}{opp}$

$\cos\theta=\dfrac{adj}{hyp}$ $\sec\theta=\dfrac{hyp}{adj}$

$\tan\theta=\dfrac{opp}{adj}$ $\cot\theta=\dfrac{adj}{opp}$

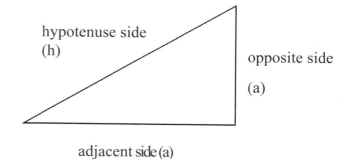

hypotenuse side (h)

opposite side (a)

adjacent side (a)

Evaluate each of the six trigonometric ratios of θ (assume θ is in first Quadrant)

1. Given: $\sin\theta = 7/10$

2. Given: $\tan\theta = 2$

3. Given: $\csc\theta = 13/12$

4. The given point (-3, -4) is on the terminal side of an angle in standard position. Determine the exact values of the six trigonometric functions of the angle.

5. The given point (5, -13) is on the terminal side of an angle in standard position. Determine the exact values of the six trigonometric functions of the angle.

45 – 45 – 90 Triangles

30 – 60 – 90 Triangles

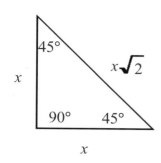

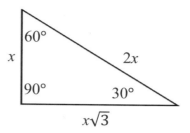

6. Use these to help develop the six trigonometric ratios for 0°, 30°, 45°, 60°, and 90°.
Assume the hypotenuse of these triangles are equal to 1.

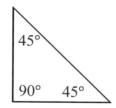

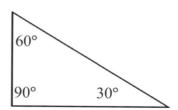

*Fill in the chart and Memorize all !!

θ	0°	30°	45°	60°	90°
sin					
cos					
tan					
cot					
sec					
csc					

Co-functions of complementary angles are equal.

$\sin(90 - \theta) = \cos\theta$ $\cos(90 - \theta) = \sin\theta$ $\tan(90 - \theta) = \cot\theta$

$\cot(90 - \theta) = \tan\theta$ $\sec(90 - \theta) = \csc\theta$ $\csc(90 - \theta) = \sec\theta$

Fundamental Trigonometric Identities

Reciprocal Identities:

$\sin\theta = \dfrac{1}{\csc\theta}$ $\csc\theta = \dfrac{1}{\sin\theta}$

$\cos\theta = \dfrac{1}{\sec\theta}$ $\sec\theta = \dfrac{1}{\cos\theta}$

$\tan\theta = \dfrac{1}{\cot\theta}$ $\cot\theta = \dfrac{1}{\tan\theta}$

Quotient Identities:

$\tan\theta = \dfrac{\sin\theta}{\cos\theta}$

$\cot\theta = \dfrac{\cos\theta}{\sin\theta}$

Pythagorean Identities:

$\sin^2\theta + \cos^2\theta = 1$

$1 + \tan^2\theta = \sec^2\theta$

$1 + \cot^2\theta = \csc^2\theta$

7. Find the value of each trig function using the trig identities when $\tan\theta = \sqrt{3}$ (θ is in 1st Quadrant)

a.) $\sec\theta =$ _____   b.) $\cos\theta =$ _____   c.) $\sin\theta =$ _____

d.) $\cot\theta =$ e.) $\csc\theta =$

8. Use the trig identities to transform one side of the equation into the other:

a.) $\csc\theta \tan\theta = \sec\theta$

b.) $\cot\alpha \sin\alpha = \cos\alpha$

c.) $\dfrac{\csc A}{\cot A + \tan A} = \cos A$

9. According to the safety sticker on a 30-foot ladder, the distance from the bottom of the ladder to the base of the wall on which it leans should be 35% of the length of the ladder.

a.) If the ladder is in this position, what is the acute angle between the bottom of the ladder and the ground?

b.) How high up the wall will the ladder reach?

Lecture 4.4 Trigonometric Functions of Any Angle

Let θ be an angle in standard position with (x,y), a point on the terminal side of the angle and $r = \sqrt{x^2 + y^2}$. $r \neq 0, \ x \neq 0 \ y \neq 0$

$$\sin\theta = \frac{y}{r} \qquad \cos\theta = \frac{x}{r} \qquad \tan\theta = \frac{y}{x}$$

$$\csc\theta = \frac{r}{y} \qquad \sec\theta = \frac{r}{x} \qquad \cot\theta = \frac{x}{y}$$

Note: If x = 0, then tan θ and sec θ are undefined.
　　　　If y = 0, then cot θ and csc θ are undefined.

1. Let (-5, 12) be a point on the terminal side of θ. Find all six trig ratios.

Trigonometry ASTC

When you work with trigonometry, you'll be dealing with four quadrants of a graph. The x and y axis divides up a coordinate plane into four separate sections.

ASTC is a memory-aid for memorizing whether a trigonometric ratio is positive or negative in each quadrant: [Add-Sugar-To-Coffee]

If you don't like Add Sugar To Coffee, there's other acronyms you can use such as:

All students Take Calculus.

	Quadrant 1	Quadrant 2	Quadrant 3	Quadrant 4
sine& Cosecant				
Cosine& Secant				
Tangent & Cotangent				

2. Given: Tan θ = -8/15 and cos θ > 0. Find the remaining trig ratios.

3. Fill in the following chart:

	Sin θ	Cos θ	Tan θ	Cot θ	Sec θ	Csc θ
π/2						
π						
3π/2						
2π or 0						

Reference Angle

The angle that the given angle makes with the x-axis. Regardless of where the angle ends (that is, regardless of the location of the terminal side of the angle), the reference angle measures the closest distance of that terminal side to the x-axis

4. Find the reference angle for each of these:

a.) 110°　　　　　　b.) 250°　　　　　　c.) - 315°　　　　d.) 520°

e.) 4π/3　　　　　　f.) 3 π/5　　　　　　g.)　13π/6

5. In which quadrants are the trig ratios positive?

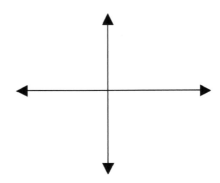

6. Evaluate the trigonometric functions for each of the following: (exact answers only)

a.) $\cos(5\pi/3)$

b.) $\tan(-240°)$

c.) $\csc 9\pi/4$

7. Let θ be an angle in Quadrant II such that $\sin\theta = 1/4$. Using trig identities, find the following:

a.) $\csc\theta$

b.) $\tan\theta$

c.) $\cos\theta$

d.) $\sin\theta \cdot \cot\theta$

8. Use your calculator to evaluate each of the following: Round values to four decimal places:

a.) $\tan(-200°)$

b.) $\cos 112°$

c.) $\csc(-247°)$

d.) $\tan(4\pi/3)$

e.) $\sin(12\pi/7)$

f.) $\sec(15\pi/4)$

9. Use your calculator to find <u>two</u> approximate values of θ $(0° \leq \theta \leq 360°)$. Round your answers to two decimal places.

a.) $\cos \theta = 0.7820$ b.) $\sin \theta = 0.3880$ c.) $\tan \theta = 1.7693$

10. Find the indicated trigonometric value in the specified quadrant:

a.) $\csc \theta = -2$, Quadrant IV, find $\sec \theta$.

b.) $\sec \theta = -7/4$, Quadrant III, find $\cot \theta$.

c.) $\sin \theta = \frac{2}{5}$, Quadrant II, find $\tan \theta$.

Lecture 4. 5 Graph of Sine and Cosine

1. Fill in the chart below.

X	0	$\dfrac{\pi}{4}$	$\dfrac{\pi}{2}$	$\dfrac{3\pi}{4}$	π	$\dfrac{5\pi}{4}$	$\dfrac{3\pi}{2}$	$\dfrac{7\pi}{4}$	2π
$y=\sin x$									
$y=\cos x$									

2. Using the table above to sketch a graph of the sine and cosine functions.

a.) $y = \sin x$

Amplitude=

Period=

Starts at

b.) $y = \cos x$

Amplitude=

Period=

Starts at

Sin and Cos Graph Key Concept

$y = A \sin[b(x - c)] + d$ $\qquad\qquad$ $y = A \cos[b(x - c)] + d$

Amplitude $= |A|$ $\qquad\qquad$ Period $= \dfrac{2\pi}{b}$

Amplitude $= \dfrac{\text{Max} - \text{Min}}{2}$

d represents a vertical shift of the sine or cosine curve.

Phase Shift: (set the quantity inside the parenthesis equal to zero and solve For x (c in the cases above).

3. Graph, on the same coordinate axis, one period of each of the following functions.

a.) $y = sin(x)$ $\qquad\qquad$ b.) $y = 2 \sin (x)$ $\qquad\qquad$ c.) $y = \frac{1}{2}sin(x)$

4. Graph one period of each of the following functions.

a.) $y = sin3x$ $\qquad\qquad\qquad$ b.) $y = cos\left(\frac{1}{3}x\right)$

c.) $y = \sin\left(x - \dfrac{\pi}{2}\right)$

d.) $y = -\cos\left(\dfrac{\pi}{3}x\right)$

5. Find the period and amplitude for each of the following functions.

a.) $f(x) = 3\sin\left(\dfrac{\pi}{2}x\right)$

b.) $g(x) = 3 + 5\sin\left(\dfrac{\pi}{2}x - \dfrac{\pi}{2}\right)$

c.) $h(x) = \dfrac{7}{3}\cos\left(\dfrac{3}{2}x\right)$

6. Describe the relationship between the graphs f and g.

a.) $f(x) = \cos x$

$g(x) = -\cos\left(x + \dfrac{\pi}{4}\right)$

b.) $f(x) = 2\cos x$

$g(x) = -3 + 2\cos\left(x - \dfrac{\pi}{2}\right)$

c.) $f(x) = \sin 3x$

$g(x) = 2 + \sin 3x$

7. Graph one period of each of the following functions

a.) $f(x) = 2 + 2\sin\left(\frac{x-\pi}{3}\right)$

b.) $g(x) = -4 + \cos\left(2x - \dfrac{\pi}{4}\right)$

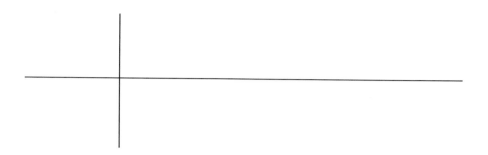

c.) $f(x) = \pi \, \sin\left(\dfrac{x+1}{2\pi}\right) + 3$

Lecture 4.6 Graph of Tangent and Cotangent

1. Fill in the chart below.

x	o	$\dfrac{\pi}{4}$	$\dfrac{\pi}{2}$	$\dfrac{3\pi}{4}$	π	$\dfrac{5\pi}{4}$	$\dfrac{3\pi}{2}$	$\dfrac{7\pi}{4}$	2π
$y=\tan x$									
$y=\cot x$									

2. Graph the following:

a.) $y = \tan x$

Period=

Starts at _____

b.) $y = \cot x$

Period=

Starts at ____

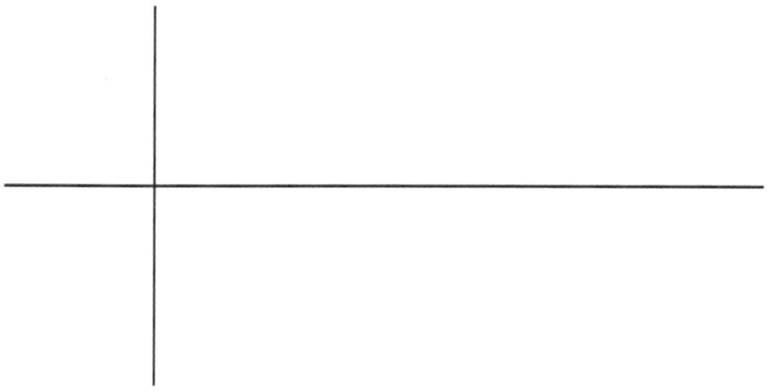

Tan. and Cot. Graph Key Concept

y =A tan[b(x-c)]+d

y=A cot[b(x-c)]+d

A: describes steepness of curve

Period=$\frac{\pi}{b}$

d represents a vertical shift of the sine or cosine curve

Phase Shift (set the quantity inside the parenthesis equal to zero and solve
For x (c in the cases above)

3. Give the period and graph one cycle of each of the following functions.

a.) $y=\tan(\frac{1}{2}x)$

b.) $y=2\cot\pi x$

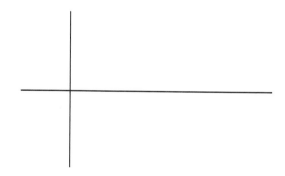

c.) $y=\cot(\pi x + \pi)$

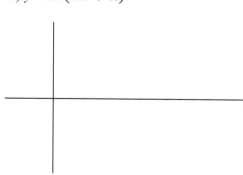

d.) $y=2\cot\left(x + \frac{\pi}{2}\right)+1$

4. Graphing the sec x and csc x (First graph the sin x and the cos x)

y=csc x

y=sec x

5. Graph one cycle of each of the following:

a.) $y=3\csc(2x)+1$

b.) $y=-2\sec[\frac{\pi}{4}(x-1)]$

c.) $y=2+2\csc\left(x+\frac{\pi}{4}\right)$

d.) $y=\frac{1}{2}\sec\left(\frac{\pi}{2}x+\frac{\pi}{2}\right)-3$

Lecture 4.7 Inverse Trigonometric Functions

Inverse Trigonometric Functions

For a function to have an inverse, it must pass the horizontal line test or we must restrict the domain, which then restricts the range of the inverse

$y = \arcsin x$ **is the same as** $y = \sin^{-1}x$

1. Examine the graphs of y=sinx and y=arcsinx. (Use your calculators to draw the inverses.) Then draw the inverse like we did above.

y=sinx   y=arcsinx

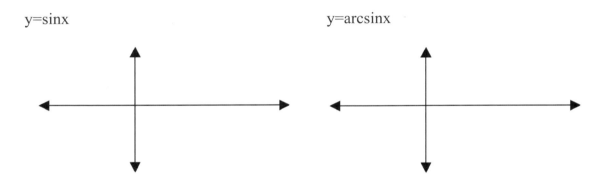

Function	Range
y=arcsin x	$\dfrac{-\pi}{2} \leq y \leq \dfrac{\pi}{2}$
y=arccos x	$0 \leq y \leq \pi$
y=arctan x	$\dfrac{-\pi}{2} < y < \dfrac{\pi}{2}$

Sketch a graph of

$y = \arctan x$ y=arccos x

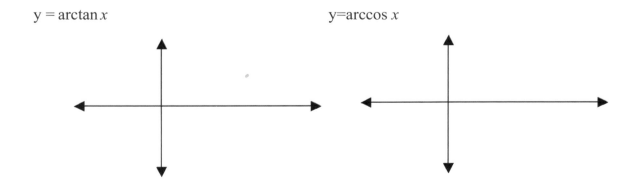

2. Without Using a calculator, find the exact values of each expression ($0° \leq \theta \leq 360°$).:

a.) $\arcsin\left(\frac{-\sqrt{3}}{2}\right)$

b.) $\arccos\left(\frac{-\sqrt{2}}{2}\right)$

c.) $\arctan\left(\sqrt{3}\right)$

d.) $\arccos\left(\frac{1}{2}\right)$

3. Using a calculator, find the exact values of each expression ($0° \leq \theta \leq 360°$).:

a.) $\arcsin(-0.125)$

b.) $\arctan(17.8)$

c.) arcsec (2.1)

d.) arccsc (-3.5)

Note: $\sin(\arcsin x) = x$ and $\arcsin(\sin y) = y$ (Same with cos, tan), assuming x and y fall within the restrictions.

4. Evaluate ($0° \leq \theta \leq 360°$):

a.) $\sin(\arcsin 0.7) =$

b.) $\cos(\arccos(-0.3)) =$

5. Without using a calculator, find the exact value of the expression ($0° \leq \theta \leq 360°$).

a.) $\arcsin\left(\sin\left(\frac{\pi}{6}\right)\right) =$

b.) $\cos\left(\arcsin\left(\frac{-1}{3}\right)\right) =$

c.) $\cot\left(\arcsin\left(\frac{\sqrt{3}}{2}\right)\right) =$

d.) $\cos\left(\arcsin\left(-\frac{4}{5}\right)\right) =$

6. Write an algebraic expression that is equivalent to the expression ($0° \leq \theta \leq 360°$):

a.) $\sin(\arccos x) =$

b.) $\cos(\arcsin(3x)) =$

c.) $\sec\left(\arccos\left(\frac{1}{x}\right)\right) =$

5

Advance Trigonometry

Learning Objective

Know how to Prove and solve complicated trig. Equations using Trig. Identities.

Lecture 5.1 Fundamental Trig. Identities

Important Trig. Identities

Reciprocal Identities:

$$\sin \theta = \frac{1}{\csc \theta} \qquad \cos \theta = \frac{1}{\sec \theta} \qquad \tan \theta = \frac{1}{\cot \theta}$$

$$\csc u = \frac{1}{\sin u} \qquad \sec u = \frac{1}{\cos u} \qquad \cot u = \frac{1}{\tan u}$$

Quotient Identities:

$$\tan u = \frac{\sin u}{\cos u} \qquad \cot u = \frac{\cos u}{\sin u}$$

Pythagorean Identities:

$$\sin^2 u + \cos^2 u = 1 \qquad 1+\tan^2 u = \sec^2 u \qquad 1+\cot^2 u = \csc^2 u$$

$$\cos^2 u = 1 - \sin^2 u$$

$$\sin^2 u = 1 - \cos^2 u$$

Cofunction Identities:

$$\sin \left(\frac{\pi}{2} - u\right) = \cos u \qquad \cos \left(\frac{\pi}{2} - u\right) = \sin u \qquad \tan \left(\frac{\pi}{2} - u\right) = \cot u$$

$$\cot \left(\frac{\pi}{2} - u\right) = \tan u \qquad \sec \left(\frac{\pi}{2} - u\right) = \csc u \qquad \csc \left(\frac{\pi}{2} - u\right) = \sec u$$

Even/Odd Identities:

$$\sin(-u) = -\sin u \qquad\qquad \cos(-u) = \cos u$$

$$\tan(-u) = -\tan u$$

You can use above identities to:

1.) Evaluate trigonometric functions.
2.) Simplify trigonometric expressions.
3.) Solve trigonometric equations.

1. Solve $\sin x = \sqrt{3} \cos x$ for all values of x such that $0 \leq x < 2\pi$.

2. Use the values of $\sec \theta = \frac{-5}{2}$ and $\tan \theta > 0$ to find the values of all six trigonometric functions.

3. Simplify: $\sin x \cdot \cos^2 x - \sin x$

4. Simplify: $\sin x + \cot x \cdot \cos x$

5. Factor: $\csc^2 x - \cot x - 3$

6. Add and Simplify: $\dfrac{\sin \theta}{1+\cos \theta} + \dfrac{\cos \theta}{\sin \theta}$

7. Rewrite $\dfrac{1}{1+\sin x}$ so that it is not in fractional form.

8. Simplify: $\dfrac{\sin\theta+\cos\theta}{\sin\theta}+\dfrac{-\cos\theta+\sin\theta}{\cos\theta}$

9. Expand and Simplify: $(\cot x + \csc x)(\cot x - \csc x)$

"Shortcut to Reduction Formulas" using Co-Functions (Optional)

Regular Functions		Co-functions	Rules
$\sin x$ Function	$\leftrightarrow$	$\cos x$	1.) Horizontal -> Same
$\tan x$	$\leftrightarrow$	$\cot x$	2.) Vertical -> Use Co-function
$\sec x$	$\leftrightarrow$	$\csc x$	

Use the shortcut method: (Assume $< \frac{\pi}{2}$)

10. Simplify: $\sin\left(\frac{\pi}{2} + x\right) =$

11. Simplify: $\csc\left(\frac{3\pi}{2} - x\right) =$

12. Simplify: $\tan(\pi - x) =$

13. Simplify: $\cos(2\pi - x) =$

Lecture 5.2 Proving and Verifying trigonometric identities

Strategies

1.) Work with only left side of the equation.
2.) Factor

3.) Substitute proper identities
4.) Change everything to sin x and cos x and simplify

Verify the following identities:

1. $\dfrac{\sin^2 x - 1}{\sin^2 x} = -\cot^2 x$

2. $\dfrac{1}{\sec x - 1} - \dfrac{1}{\sec x + 1} = 2\cot^2 x$

3. $(1 + \cot^2 x)(1 - \sin^2 x) = \cot^2 x$

4. $\sec u + \tan u = \dfrac{1}{\sec u - \tan u}$

5. $\dfrac{\cot^2 \theta}{1 + \csc \theta} = \dfrac{1 - \sin \theta}{\sin \theta}$

6. $\cot t \cos t = \dfrac{1}{\tan t \sec t}$

7. $(\tan^2 x + 1)(\cos^2 x - 1) = -\tan^2 x$

8. $\dfrac{\cos x}{1-\sin x} = \sec x + \tan x$

Lecture 5.3 Solving Trigonometric Equations

Solve the following trigonometric equations: (Isolate the trig function)

1. $2 - 4\cos x = 0$

Solutions in the interval: $[0, 2\pi)$

All Solutions possible (General):

2. $\sin x + 2 = -3\sin x$

Solutions in the interval: $[0.2\pi)$

All Solutions possible (General):

3. Solve: $2\tan^2 x - 6 = 0$ in the interval: $[0.2\pi)$

4. Solve: $\sec x \csc x = \csc x$ in the interval: $[0.2\pi)$

5. Solve: $2\cos^2 x + \cos x - 1 = 0$ in the interval: $[0.2\pi)$

6. Solve: $2\cos^2 x + 3 \sin x - 3 = 0$ in the interval: $[0.2\pi)$

7. Solve: $\cos x + 1 = \sin x$ in the interval: $[0.2\pi)$

8. solve: $2 \sin 2t + 1 = 0$ in the interval: $[0.2\pi)$

9. Solve: $2 \cos 3\theta - 1 = 0$ in the interval: $[0.2\pi)$

Lecture 5.4 Sum and Difference Formulas

$$\sin(u + v) = \sin u \cos v + \cos u \sin v \qquad \sin(u - v) = \sin u \cos v - \cos u \sin v$$

$$\cos(u + v) = \cos u \cos v - \sin u \sin v \qquad \cos(u - v) = \cos u \cos v + \sin u \sin v$$

$$\tan(u + v) = \frac{\tan u + \tan v}{1 - \tan u \tan v} \qquad \tan(u - v) = \frac{\tan u - \tan v}{1 + \tan u \tan v}$$

1. Find the exact value of $\sin 75°$.

2. Find the exact value of $\cos \frac{7\pi}{12}$.

3. Find the exact value of $\cos 58° \cos 13° + \sin 58° \sin 13°$.

4. Find the exact value of $\tan 195°$.

5. Find the exact value of $\sin (x + y)$ given that $\sin x = \frac{3}{5}$, where $0 < x < \frac{\pi}{2}$ and $\cos y = \frac{-5}{13}$, where $\frac{\pi}{2} < y < \pi$.

6. If A is an angle in the third quadrant, B is an angle in the second quadrant, $\tan A = \frac{3}{4}$, and $\tan B = -\frac{1}{2}$, in which quadrant does angle $(A + B)$ lie?

7. Simplify each expression using the addition formulas:

a.) $\sin(90° - x) =$

b.) $\cos(180° + x) =$

8. Find all solutions of $\sin\left(x + \frac{\pi}{4}\right) + \sin\left(x - \frac{\pi}{4}\right) = -1$, where $0 \leq x < 2\pi$.

Lecture 5.5 Double Angle Formulas

Multiple-Angles and Product-Sum Formulas

Double Angle Formulas

$$\sin 2u = 2\sin u \cos u$$

$$\cos 2u = \cos^2 u - \sin^2 u$$

$$\tan 2u = \frac{2\tan u}{1 - \tan^2 u}$$

$$\cos 2u = 2\cos^2 u - 1$$

$$\cos 2u = 1 - 2\sin^2 u$$

Develop the double angle formulas for sine and cosine.

$\sin(2u) =$

$\cos(2u) =$

1. Express $\sin 4x$ in terms of $\sin x$ and $\cos x$.

2. Solve the equation: $\sin 2x - \cos x = 0$ in the interval $[0, 2\pi)$

3. Find the value of $\cos 2\theta - \sin (90°+\theta)$ if $\tan \theta = -\dfrac{3}{4}$ and $\sin \theta$ is positive.

4. Solve the equation: $2 \cos x + \sin 2x = 0$ in the interval $[0, 2\pi)$

5. Given $\sin x = \frac{12}{13}$ and $\frac{\pi}{2} < x < \pi$, find $\sin 2x, \cos 2x, and \tan 2x$.

Half-Angle Formulas

$$\sin\frac{u}{2} = \pm\sqrt{\frac{1-\cos u}{2}} \qquad \cos\frac{u}{2} = \pm\sqrt{\frac{1+\cos u}{2}} \qquad \tan\frac{u}{2} = \frac{1-\cos u}{\sin u} = \frac{\sin u}{1+\cos u}$$

*The signs of $\sin\left(\frac{u}{2}\right)$ and $\cos\left(\frac{u}{2}\right)$ depend on the quadrant in which $\left(\frac{u}{2}\right)$ lies.

6. Find the exact value of the $\cos 165°$ using the half-angle formula.

7. Find the exact value of the $\sin 105°$ using the half-angle formula.

8. Prove: $\dfrac{\sin t + \sin 3t}{\cos t + \cos 3t} = \tan 2t$

Product-to-Sum Formulas (For Honor classes)

$$\sin u \sin v = \frac{1}{2}[\cos(u-v) - \cos(u+v)], \cos u \cos v = \frac{1}{2}[\cos(u-v) + \cos(u+v)]$$

$$\sin u \cos v = \frac{1}{2}[\sin(u+v) + \sin(u-v)], \cos u \sin v = \frac{1}{2}[\sin(u+v) - \sin(u-v)]$$

9. Rewrite the product $\cos 5x \sin 4x$ as a sum or difference.

Sum-to-Product Formulas

$$\sin x + \sin y = 2 \sin\left(\frac{x+y}{2}\right) \cos\left(\frac{x-y}{2}\right) \qquad \sin x - \sin y = 2 \cos\left(\frac{x+y}{2}\right) \sin\left(\frac{x-y}{2}\right)$$

$$\cos x + \cos y = 2 \cos\left(\frac{x+y}{2}\right) \cos\left(\frac{x-y}{2}\right) \qquad \cos x - \cos y = -2 \sin\left(\frac{x+y}{2}\right) \sin\left(\frac{x-y}{2}\right)$$

10. Rewrite $\cos 4x + \cos 6x$ as a product.

11. Find the solutions of $\sin 5x + \sin 3x = 0$, where $0 \le x < 2\pi$.

6

Trigonometry
Additional Topics

Learning Objective

Know how to solve any type of triangles
using Law of Sine and Cosine.
Know and understand basic Vector
Concepts.

Lecture 6.1 Law of Sine

Using the Law of Sines, we can solve any types of triangles (acute, obtuse, and right)

The law of sines is most useful if the following items are given:
1.) Two angles and any side (A-A-S or A-S-A)
2.) Two sides and an angle opposite one of them (S-S-A) – (ambiguous)

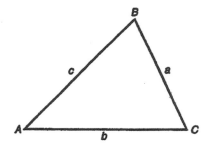

Law of Sines: If ABC is a triangle with sides a, b, and c, then,

$$\frac{a}{\sin A} = \frac{b}{\sin B} = \frac{c}{\sin C} \quad \text{which is the same as} \quad \frac{\sin A}{a} = \frac{\sin B}{b} = \frac{\sin C}{c}$$

Given the following information, find the remaining sides and angles of the triangle.

1. $\angle B = 25°$, $\angle C = 38°$ and b = 210

2. $\angle A = 50°$, $\angle B = 28°$ and a= 21

Ambiguous Case (Side – Side – Angle) S-S-A

Three possibilities can occur when ∠A < 90° :

1.) No triangles exist. $\qquad\qquad$ ($a < c \sin A$)

2.) One triangle exists. $\qquad\qquad$ ($a = c \sin A$ or $a \geq c$)

3.) Two distinct triangles exist. $\quad$ ($c \sin A < a < c$)

Examine the height of the triangle (h). How can we relate that to ∠A?

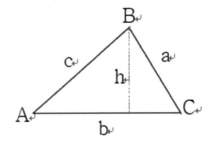

Given the following information, determine the number of triangles that exist.

3. $a = 14.8, b = 25.4$ and $∠A = 82°$ $\qquad\qquad\qquad$ <u>Drawing</u>

4. $a = 13, b = 21$ and $∠A = 43°$ $\qquad\qquad\qquad$ <u>Drawing</u>

5. $a = 49, b = 35$ and $\angle B = 60°$ <u>Drawing</u>

<u>Areas of Oblique Triangles (Develop Area Formulas)</u>

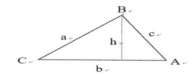 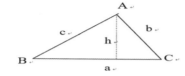 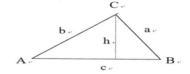

Area of a Δ:　Area $= \dfrac{1}{2} bc \cdot \sin A$

Area $= \dfrac{1}{2} ac \cdot \sin B$

Area $= \dfrac{1}{2} ab \cdot \sin C$

6. Find the area of ΔABC if $a = 180$ inches, $b = 150$ inches, and $\angle C = 30°$.

7. Find the area of a triangular lot having two sides of 7.5 feet and 13 feet and the included angle of $120°$.

Lecture 6.2 Law of Cosine

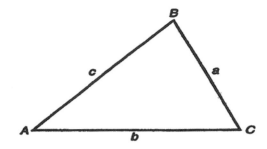

Using the Law of Cosines, we can solve all types of triangle including oblique triangles

The law of cosine is most useful if the following items are given:
1.) all 3 sides: Side – Side – Side (SSS)
2.) 2 sides and 1 adjacent side: Side – Angle – Side (SAS)

Law of Cosines

<u>Standard Form</u>   <u>Alternative Form</u>

$$a^2 = b^2 + c^2 - 2bc\cos A$$

$$b^2 = a^2 + c^2 - 2ac\cos B$$

$$c^2 = a^2 + b^2 - 2ab\cos C$$

$$\cos A = \frac{b^2 + c^2 - a^2}{2bc}$$

$$\cos B = \frac{a^2 + c^2 - b^2}{2ac}$$

$$\cos C = \frac{a^2 + b^2 - c^2}{2ab}$$

General Rule:

Start by finding the largest angle of the triangle first.

Solve the following triangles:

1. In triangle ABC, $a = 10, b = 6$, and $\angle C = 25°$.

2. In triangle ABC, $a = 12, b = 8,$ and $c = 6$.

Heron's Area Formula (Used when given S-S-S)

Given any triangle with sides a, b, and c, the area of the triangle is:

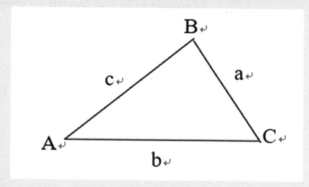

$$\text{Area} = \sqrt{s(s-a)(s-b)(s-c)}$$ **where** $s = \frac{a+b+c}{2}$

3. Find the area of a triangle with sides 13 inches, 15 inches, and 18 inches.

4. A surveyor finds that the edges of a triangular lot measure 42.5m, 37.0m, and 28.5m. What is the area of the lot?

Lecture 6.3 Basic Vector

Vectors in the Plane

Vector-A directed line segment that represents both magnitude and direction.

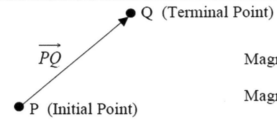

Q (Terminal Point)

$\overrightarrow{PQ}$

P (Initial Point)

Magnitude of $\overline{PQ}$ is represented as $|\overrightarrow{(PQ)}|$

Magnitude is the length of a vector.

1. Suppose we have vector $\left|\overrightarrow{PQ}\right|$, with $P = (1,0)$ and $Q = (5,3)$. Draw the vector and find the magnitude of the vector. (length of vector)

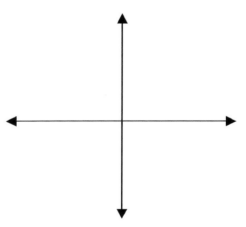

Magnitude=_____

Now, move the vector so that it starts at (0,0).
Where would the terminal point now be?

When a vector is repositioned so that it starts at (0,0), it is in component form or, sometimes, we call this a standard form.

Component Form = Terminal Point − Initial Point

2. Find the component form of vector $\overrightarrow{ab}$, if a= $(-4, 9)$ and b= $(-5, -8)$.

3. What is $|\overrightarrow{ab}|$?

Once a vector is in component form, written as $v = \langle v_1, v_2 \rangle$, then $\|v\| = \sqrt{v_1^2 + v_2^2}$.

4. Suppose $v = \langle 3, 7 \rangle$, Find $\|v\|$.

5. Given the vectors u and v. Draw the following:

a.) u + v   b.) 2u

c.) u − v   d.) v + .5u

6. Let $v = \langle -6, 4 \rangle$ and $w = \langle 2, 5 \rangle$. Find the following algebraically:

a.) $2v$　　　　　　　　　b.) $w - v$　　　　　　　　　c.) $2v + w$

A Unit Vector is a vector of magnitude = 1

A unit vector u in the direction of $v = \dfrac{v}{\|v\|} = \dfrac{1}{\|v\|}v$

---

7. Find a unit vector u in the direction of $v = \langle 3, -2 \rangle$.

Standard unit vectors are $i = \langle 1, 0 \rangle$ and $j = \langle 0, 1 \rangle$ [$k = \langle 0, 0, 1 \rangle$ *if 3 Dimensional*]

8. Let u be the vector with initial point (3, -7) and terminal point (4,9). Write u as a linear combination of the standard unit vectors i and j.

9. Suppose u=-2i+5j and v=i-7j. Find 3u+4v.

Direction of a Vector

Suppose $u = \langle a, b \rangle$ then the $\tan\theta = \frac{b}{a}$, where θ is the angle that the vector makes with the positive x-axis.

10. Find the direction angle of each vector. (Check quadrants)

a.) $u = 2i + 2j$

b.) $u = -2i - 5j$

11. Find the component form of the vector that represents the velocity of an airplane descending at a speed of 90 miles per hour at an angle of $210°$.

12. Find the vector v with the given magnitude: $\|v\| = 12$ in the direction of $u = 2i - j$

7

Sequences and Series

Learning Objective

Know how to use Arithmetic and Geometric
Sequence and Series.
Know and understand Infinite Series and
Binomial Theorem.

Lecture 7.1 Sequences and Series

An <u>infinite sequence</u> goes on forever and is written as $a_1, a_2, a_3, a_4, \cdots, a_n, \cdots$

A <u>finite sequence</u> ends and is written as $a_1, a_2, a_3, a_4, \cdots, a_n$

1. Find the first four terms of the following sequences:

a.) $a_n = 3n + 4$

b.) $b_n = \dfrac{(-1)^n \cdot 2}{n^2 + 1}$

2. Write an expression for the apparent n^{th} term of each sequence.

a.) $2, 6, 10, 14, 18, \ldots a_n$

b.) $3, 6, 11, 18, 27, \ldots a_n$

Recursive sequence

You must be given at least the one term. Then subsequent terms are defined based on the term that is given.

How recursive formulas work

Recursive formulas give us two pieces of information:

1. The one term of the sequence

2. The pattern rule to get any term from the term that comes before it

3. Here is a recursive formula of the sequence 5, 7, 9,... along with the interpretation for each part.

$$\begin{cases} a_1 = 5 \leftarrow the\ first\ term \\ a_n = a_{n-1} + 2 \leftarrow add\ 2\ to\ the\ previous\ term \end{cases}$$

3. Write the first four terms of the recursively defined sequence:

a.) $a_1 = 2$, $a_{n+1} = 3a_n + 4$

b.) $a_1 = -1$, $a_{n+1} = 2(a_n)^2 + 21$

c.) $a_2 = -3$, $a_{n+1} = -a_n + 4$

Definition of Factorial

If n is a positive integer, n factorial is defined by

$$n! = n \bullet (n - 1) \dots 3 \bullet 2 \bullet 1$$

Also note, zero factorial is defined as $0! = 1$

4. Find the following factorials:

$0! =$ $1! =$ $2! =$ $3! =$

$4! =$ $5! =$ $6! =$ $7! =$

5. Write the first four terms of the following sequence: $a_n = \dfrac{3n}{n!}$

6. Evaluate the factorial expressions:

a.) $\dfrac{9!}{3!8!}$

b.) $\dfrac{2n!}{(n+1)!}$

c.) $\dfrac{2(n-2)!}{(n-3)!}$

d.) $\dfrac{(n+1)!}{(n+5)!}$

Definition of Summation Notation:

The sum of the first n terms of a sequence is represented by

$$\sum_{i=1}^{n} a_i = a_1 + a_2 + a_3 + a_4 + \text{L} + a_n$$

where i is called the index of summation, n is the upper limit summation, and 1 is the lower limit summation.

7. Find the following sums: (expand and then simplify)

a.) $\displaystyle\sum_{i=1}^{5} 4i - 2 =$

b.) $\displaystyle\sum_{i=2}^{7} (-1)^{i-1} \cdot i! =$

Properties of Sums

1.) $\displaystyle\sum_{i=1}^{n} ca_i = c\sum_{i=1}^{n} a_i$ 2.) $\displaystyle\sum_{i=1}^{n}(a_i + b_i) = \sum_{i=1}^{n} a_i + \sum_{i=1}^{n} b_i$ 3.) $\displaystyle\sum_{i=1}^{n}(a_i - b_i) = \sum_{i=1}^{n} a_i - \sum_{i=1}^{n} b_i$

An infinite series is the sum of the terms of an infinite sequence and it is written as:

$$a_1 + a_2 + a_3 + \cdots + a_i + \cdots = \sum_{i=1}^{\infty} a_i \quad \text{(infinite series)}$$

A finite sequence is the sum of the terms of a finite sequence and it is written as:

$$a_1 + a_2 + a_3 + \cdots + a_n = \sum_{i=1}^{n} a_i \quad \text{(finite series)}$$

8. For the series $2 \cdot \displaystyle\sum_{i=1}^{\infty} \frac{3}{2^i}$, find (a) and (b)

a.) The third partial sum is:

b.) The approximate final sum would be:

9. Use sigma notation to write the sum.

a.) $1 + 4 + 7 + 10 + 13 + 16 + 19 + 22$

b.) $\frac{2}{1} + \frac{2}{2} + \frac{2}{6} + \frac{2}{24} + \frac{2}{120}$

10. Find the sum of $\displaystyle\sum_{k=0}^{4}(-2k + k!)$.

Lecture 7.2 Arithmetic Sequences and Partial Sums

Arithmetic sequence

a sequence in which the difference between consecutive terms is always the constant. (d would be the common difference)

$$a_n = a_1 + (n-1)d \ \text{ or } \ a_n = dn + c$$

a_n is the nth term, a_1 is the 1st term, n is the term number, d is the common difference, and c is the shift of sequence

To find the common difference, d, subtract to consecutive terms:

$$d = a_n - a_{n-1}$$

1. Find the common difference in the following arithmetic sequences:

a.) $3, 8, 13, 15, ...$

b.) $25, 11, -3, -17, ...$

c.) $1, \frac{2}{3}, \frac{1}{3}, 0, \frac{-1}{3}, ...$

2. Determine if the sequence is arithmetic:

a.) $2, 6, 10, 4, 10, 12, ...$

b.) $1, -4, -9, -14, ...$

c.) $1, \frac{1}{2}, \frac{1}{3}, \frac{1}{4}, ...$

3. Find the 6th term of the arithmetic sequence with common difference 5 and first term 9.

4. Find the nth term of the arithmetic sequence with fifth term 19 and ninth term 27.

5. Find the ninth term of the arithmetic sequence whose first two terms are 1 and 6.

The sum of a finite arithmetic sequence with n terms is

$$S_n = \frac{n}{2}(a_1 + a_n) \qquad \text{or} \qquad S_n = \frac{n(2a_1 + (n-1)d)}{2}$$

6. Find the sum: $1 + 6 + 11 + 16 + 21 + 26 + 31 + 36 + 41 + 46 + 51$ (Use the formula)

7. Find the sum of the first 12 multiples of 4.

8. Find the 15th partial sum of the sequence $2, 5, 8, 11, \ldots$

9. Find the sum: $\displaystyle\sum_{n=1}^{100} 2n =$

10. Find the sum of the numbers from 1 to 1000.

11. Find the sum: $\displaystyle\sum_{n=3}^{45} 3n + 5 =$

Lecture 7.3 Geometric Sequences and Series

Geometric Sequences

A sequence is a geometric sequence if the ratios of consecutive terms are the constant.

$\frac{a_2}{a_1} = \frac{a_3}{a_2} = \frac{a_4}{a_3} = \cdots = r, r \neq 0$ (The number r is the common ratio of the sequence.)

The nth term of a geometric sequence has the form $a_n = a_1 r^{n-1}$, where r is the common ratio of consecutive terms of the sequence.

1. Find the common ratio of the following geometric sequence:

$2, 6, 18, 54$

2. Determine whether the sequence is geometric. If it is, find the common ratio.

a.) $1, -2, 4, -8, \ldots$

b.) $9, -6, 4, \frac{-8}{3}, \ldots$

c.) $\frac{1}{5}, \frac{2}{7}, \frac{3}{9}, \frac{4}{11}, \ldots$

3. Write the first four terms of the geometric sequence in which $a_1 = 9$ and $r = \frac{-1}{3}$.

4. Write the first four terms of the geometric sequence in which $a_1 = 243$ and $a_{k+1} = \frac{1}{3} a_k$.

5. Find the fifth term of a geometric sequence in which $a_1 = 3$ and $r = \frac{2}{3}$.

6. Find the twenty-2nd term of the geometric sequence $1, 3, 9, 27, ...$

Geometric Series

The sum of the geometric sequence $a_1, a_1 r, a_1 r^2, a_1 r^3, a_1 r^4, ..., a_1 r^{n-1}$ with common ratio $r \neq 1$ is

$$S_n = a_1 \left(\frac{1-r^n}{1-r} \right)$$ or $$S_n = \frac{(a_1(1-r^n))}{(1-r)}$$

7. Find the sum of $\sum_{n=2}^{11} 4(.2)^n$

8. Find the sum of $\displaystyle\sum_{i=1}^{6} 64 \left(\frac{1}{4}\right)^{i-1}$

The Sum of an Infinite Geometric Series

Exploration:

Suppose we have a sequence in which $a_1 = 2$ and the sequence went on forever (infinite). Could we find a finite sum for the infinite series? Let's look at a few of cases:

Do the following series approach a specific value on your calculator?

a.) r = 1, $2 + 2 + 2 + 2 + 2 + \cdots$

b.) r = -1, $2 - 2 + 2 - 2 + 2 + \cdots$

c.) r = 2, $2 + 4 + 8 + 16 + 32 + \cdots$

d.) $r = \frac{1}{2}$, $2 + 1 + \frac{1}{2} + \frac{1}{4} + \frac{1}{8} + \cdots$

Infinite Geometric Series

In General, if $|r| < 1$, then the infinite geometric series

$a_1, a_1r, a_1r^2, a_1r^3, a_1r^4, ..., a_1r^{n-1}, ...$ has the sum $S = \frac{a_1}{1-r}$

9. Evaluate $\displaystyle\sum_{n=1}^{\infty} 4(0.4)^{n-1}$

10. Find the total sum of $3 + 0.3 + 0.03 + 0.003 + 0.0003 + ...$

11. Find the total sum of $\displaystyle\sum_{n=1}^{\infty} \frac{1}{2}(2)^n$

Lecture 7.4 The Binomial Theorem

Binomial – a polynomial that has two terms. (ex: $(2x + y), (a - 3b), (c + b),$ etc.)

The Binomial Theorem is used to expand binomials to higher powers.

Let's examine powers of $(x + y)$

<u>Draw Pascal's Triangle</u>

$(x + y)^0 =$

$(x + y)^1 =$

$(x + y)^2 =$

$(x + y)^3 =$

Patterns exists, which eventually lead to The Binomial Theorem

The Binomial Theorem

i) The number of term = n, where n= exponent of the binomial term.

ii) The sum of exponents of each term of the expansion = n.

iii) In the expansion of $(x + y)^n$, $(x + y)^n = x^n + nx^{n-1}y + ... +_n C_r x^{n-r}y^r + \cdots$

$+nxy^{n-1} + y^n$

The coefficient of $x^{n-r}y^r$ is $_nC_r = \dfrac{n!}{(n-r)! \cdot r!}$.

The symbol $\dbinom{n}{r}$ is often used in place of $_nC_r$ to denote binomial coefficients.

1. Find the binomial coefficients:

a.) $_7C_4$ b.) $_{12}C_0$ c.) $\binom{12}{3}$ d.) $\binom{6}{1}$

Note in Combination

In General, it is true that $_nC_r = {_nC_{n-r}}$

2. Using the fact from above, $_{17}C_2=$ Why is this true?

3. Use the Binomial Theorem to expand and simplify: $(x - 3y)^4$.

4. Use the Binomial Theorem to expand and simplify: $(2x - 3y)^3$

5. Expand the binomial using Pascal's triangle to determine the coefficients:
$(2a - b)^4$

6. Find the coefficient a of the given term in the expansion of the binomial:

a.) $(2x^2 - 3)^{12}$, where the term is ax^{10}.

b.) $(2x + y)^{10}$, where the term is ax^2y^8.

7. Find the 4th term in the expansion of $(2x - 3y)^{12}$.

8. Find the 3rd term in the expansion of $(3x + 4y)^7$.

8

Analytic Geometry

Learning Objective

Recognize Conic Sections from their functions and graph each of the Conic Sections: a Circle, a Parabola, an Ellipse, and a Hyperbola.
Understand Parametric and Polar Functions.

Lecture 8.1 Conic Section: Parabola

Conic Section: the intersection of a plane and a right circular cone.

Definition of a parabola:

A parabola is the set of all points (x,y) in a plane that are equidistant from a fixed line, the directrix, and a fixed point, the focus, not on the line. The vertex is the midpoint between the focus and the directrix.

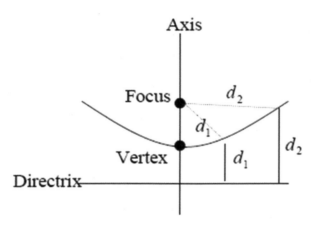

Parabola

The Standard form of the equation of a parabola with vertex (h,k) is as follows:

$(x - h)^2 = 4p(y - k), p \neq 0$ Vertical axis (up/down) directrix: y = k – p

$(y - k)^2 = 4p(x - h), p \neq 0$ Horizontal axis (left/right) directrix x = h – p

p = the distance from the vertex to the focus. It is also the distance from the vertex to the directrix.

$$(x - h)^2 = 4p(y - k)$$

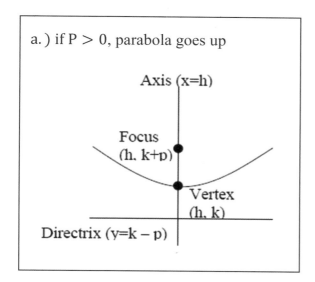

a.) if P > 0, parabola goes up

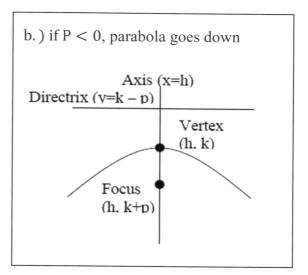

b.) if P < 0, parabola goes down

$$(y - k)^2 = 4p(x - h)$$

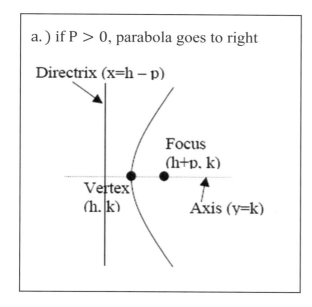

a.) if P > 0, parabola goes to right

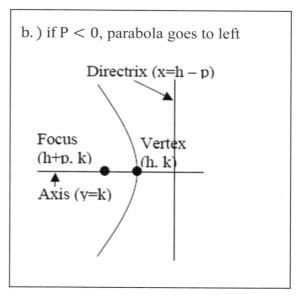

b.) if P < 0, parabola goes to left

1. Graph $y = \frac{1}{4}x^2$. Label the vertex, focus, directrix, and axis of symmetry.

2. Graph $\left(x - \frac{1}{2}\right)^2 = 2(y - 3)$. Label the vertex, focus, directrix, and axis of symmetry.

3. Graph the parabola $2y^2 - 4y - x + 5 = 0$ Label the vertex, focus, directrix, and axis of symmetry.

4. Graph the parabola: $y^2 - 6y - 4x + 17 = 0$. Label the vertex, focus, directrix, and axis of symmetry.

5. Graph the parabola: $4y^2 - 2x + 8 = 0$. Label the vertex, focus, directrix, and axis of symmetry.

6. Find the standard form of the equation of the parabola with Focus (1, -3) and Vertex (1, -2).

7. Find the standard form of the equation of the parabola with Vertex (-3, 2) and directrix of x=1.

Lecture 8.2 Ellipses

Ellipse

An ellipse is the set of all points (x, y) the sum of whose distances from two distinct fixed points (foci) is constant. $d_1 + d_2 = 2a$

$$Standard\ form: \frac{(x-h)^2}{a^2} + \frac{(y-k)^2}{b^2} = 1$$

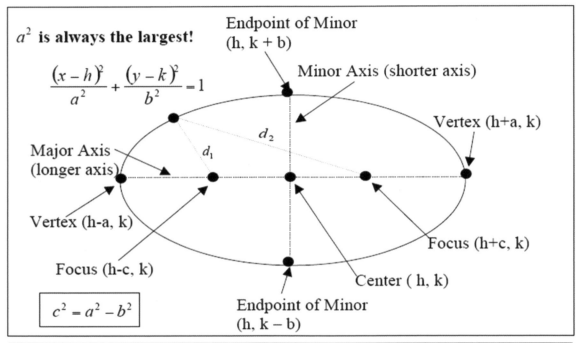

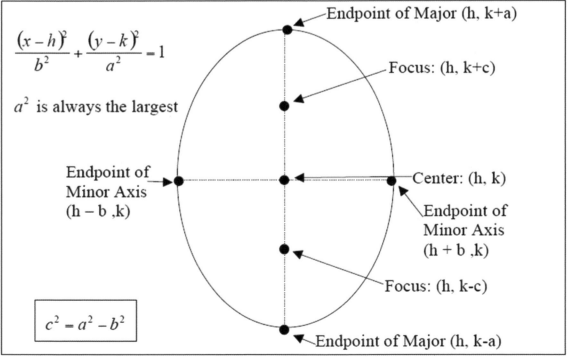

Key points for an ellipse:

1.) a^2 is always the largest denominator (a = larger stretch)
2.) $c^2 = a^2 - b^2$ and $d_1 + d_2 = 2a$ (sum of focal radii = 2a)
3.) c is the distance from the center to each of the two focus points
4.) a is the distance from the center to the vertices (endpoints of major axis)
5) b is the distance from the center to the endpoints on the minor axis (covertices)
6.) Major axis is always the <u>longest</u> axis (equals 2a)
7.) Minor axis is always the <u>shortest</u> axis (equals 2b)
8.) Before you do any operation, you must convert the equation to the standard form:
$\frac{(x-h)}{a^2} + \frac{(y-k)^2}{b^2} = 1$

1. Find the center, vertices, covertices, and foci of the ellipse given by
$\frac{x^2}{16} + \frac{y^2}{4} = 1$. Then Graph.

2. Sketch the graph of $\frac{(x-3)^2}{16} + \frac{(y+2)^2}{25} = 1$. Identify center, vertices, covertices, and foci.

3. Sketch the graph of $4x^2 + 9y^2 - 8x - 54y + 49 = 0$. Label center, vertices, covertices, and foci.

Eccentricity

Measures the ovalness of an ellipse. $eccentricity = \frac{c}{a}$.

If e is really small (close to 0), then the ellipse is almost a circle.
If e is close to 1, then the ellipse is very elongated.

The orbit of the moon has an eccentricity of e=0.0549.

4. Find the center, vertices, foci, and eccentricity of the ellipse:
$4x^2 + 9y^2 - 24x + 36y + 36 = 0$. Then, sketch its graph.

5. Find the standard form of the equation of the ellipse. Foci $(\pm 2, 1)$; Vertices $(\pm 5, 1)$

6. Find the standard form of the equation of the ellipse. Foci $(0, \pm2)$; Endpoints of the minor axis $(\pm5, 0)$

7. Find the standard form of the equation of the specified ellipse:
Foci: $(0,0),(6,0)$;   Major axis of length 10.

8. Find the standard form of the equation of the specified ellipse:
Vertices: $(\pm 4, 0)$; eccentricity of $\frac{1}{4}$.

9. Find the standard form of the equation of the specified ellipse:
Endpoints of the major axis: $(\pm 8, 0)$; Endpoints of the minor axis $(0, \pm 4)$.

Lecture 8.3 Extension (Identification of Conics)

All Conic Sections originate from the following equation

$$Ax^2 + Cy^2 + Dx + Ey + F = 0$$

1. Circle:	A=C	$(A \neq 0)$
2. Parabola:	AC=0	(A=0 or C=0 but not both)
3. Ellipse:	AC>0	(A and C have the same signs)
4. Hyperbola:	AC<0	(A and C have opposite signs)

Identify the graphs and then make a sketch. Label critical points.

1. $3y^2 - x + 1 = 0$ 2. $4x^2 - 9y^2 = 144$

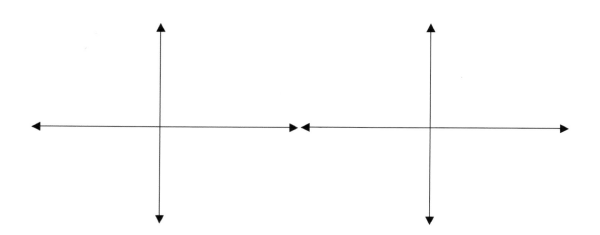

3. $x^2 + y^2 - 2y = 0$

4. $16x^2 + 9y^2 - 32x + 72y + 16 = 0$

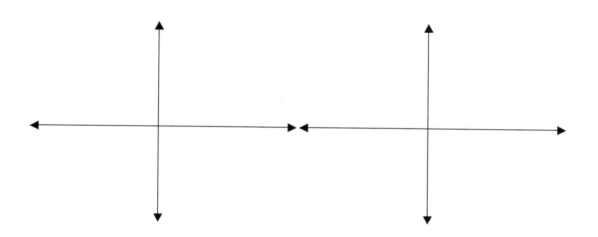

5. $-4x^2 + 25y^2 - 8x + 150y + 121 = 0$

6. $x^2 + y^2 + 6x - 8y = -9$

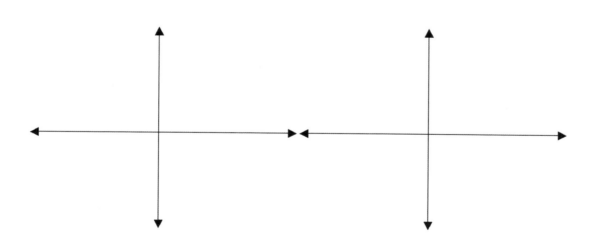

Lecture 8.4 Hyperbolas

Hyperbola

A hyperbola is the set of all points (x, y) the difference of whose distances from two distinct fixed points (foci) is a positive constant. $|d_1 - d_2| = 2a$

$$Standard\ form: \frac{(x-h)^2}{a^2} - \frac{(y-k)^2}{b^2} = 1 \quad \text{(Transverse axis is horizontal)}$$

Slopes of the asymptotes in this case is $\pm \frac{b}{a}$

$$Standard\ form: \frac{(y-k)^2}{a^2} - \frac{(x-h)^2}{b^2} = 1 \quad \text{(Transverse Axis is Vertical)}$$

Slopes of the asymptotes in this case are $\pm \frac{a}{b}$

1.) Center = (h, k)

2.) a = distance from center to each of the vertices.

3.) c = distance from center to each of the foci.

4.) b determines the slope of the asymptotes of a hyperbola.

5.) $|d_1 - d_2| = 2a$ (the difference of the focal radii equals 2a)

6.) $c^2 = a^2 + b^2$ (relationship between a, b, and c.)

7.) a^2 is the first denominator (the denominator of the positive term)

8.) If x is first, it goes left and right. If y is first it goes up and down.

9.) Transverse axis: The line segment from one vertex to the other vertex.
(the length of the transverse axis is 2a)

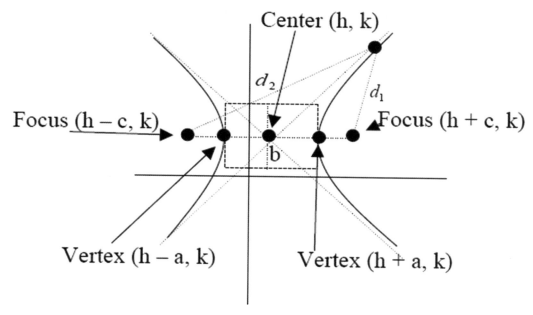

(Transverse axis is horizontal)
This occurs when x^2 is positive and y^2 is negative.

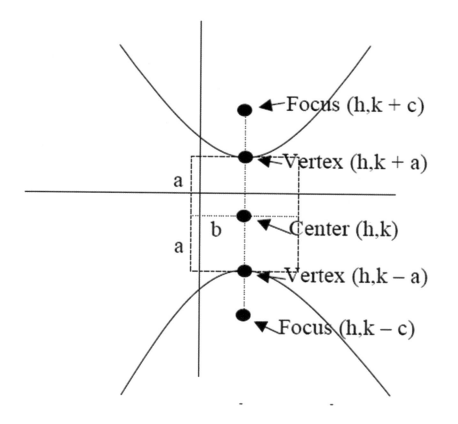

(Transverse Axis is Vertical)
This occurs when y^2 is positive and x^2 is negative.

1. Sketch the hyperbola: $\frac{(x-5)^2}{25} - \frac{(y+2)^2}{9} = 1$. Label the center, vertices, and foci.

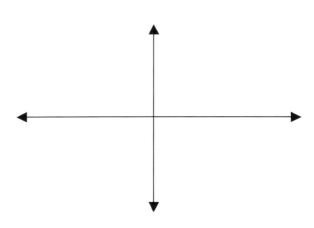

2. Sketch the hyperbola: $8y^2 - 50x^2 = 200$. Label the center, vertices, and foci.

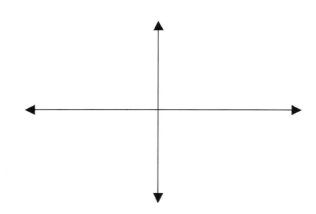

To Remember the slopes of the asymptotes, think of $slope = \frac{change\ in\ y}{change\ in\ x}$. **Then, whatever denominator that is with the y must be in the numerator.**

Example 1: $\frac{(x-3)^2}{36} - \frac{(y-5)^2}{9} = 1$;   The slope is $\pm\frac{b}{a}$, since y is second.
In this example the m= $\pm\frac{3}{6} = \pm\frac{1}{2}$.

Example 2: $\frac{(y+4)^2}{4} - \frac{(x+3)^2}{9} = 1$;   The slope is $\pm\frac{a}{b}$, since y is first.
In this example the m=$\pm\frac{2}{9}$.

Asymptotes in Hyperbola

To Write the equation of the asymptotes, first find the slope of the asymptotes, $slope = \frac{change\ in\ y}{change\ in\ x}$, then write the equation of the line through the center (h, k).

3. Given: $16y^2 - x^2 + 2x + 64y + 47 = 0$. Find the center, vertices, foci, and equations of the asymptotes. Sketch a graph.

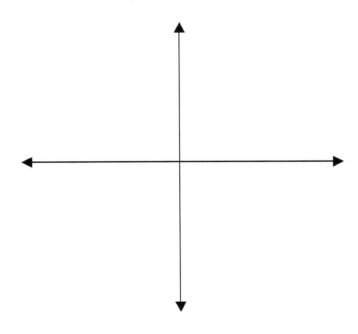

4. Find the standard form of the equation of the hyperbola with Vertices: $(\pm 3, 2)$ and Foci: $(\pm 6, 2)$

5. Find the standard form of the equation of the hyperbola with Vertices: $(0, \pm 3)$ and Asymptotes: $y = \pm 3x$.

6. Find the standard form of the equation of the hyperbola that has Vertices: (-2,1), (2,1) and passes through the point (5,4).

7. Find the standard form of the equation of the hyperbola that has Vertices: (3,0), (3, -6) and Asymptotes: $y = x - 6$ and $y = -x$.

Lecture 8. 5 Parametric Equations

Parametric Equations

a group of quantities as functions of one or more independent variables called parameters. The equations $x = f(t)$ and $y = g(t)$ are parametric equations, and t (time) is the parameter.

1. Sketch the graph of the curve given by $x = t+2$ and $y = t^2$ on $[-2, 5]$. Make a table of values and plot the points. Make sure to indicate the direction of the curve.

t	-2	-1	0	1	2	3	4	5
x								
y								

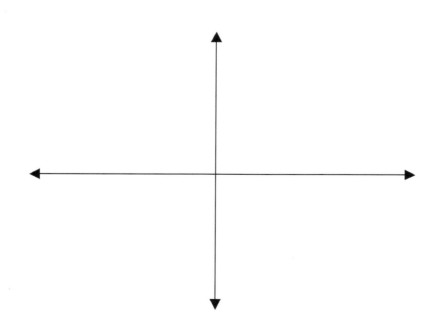

2. Use your graphing calculator to sketch the graph of the following curves on $[-2, 4]$. $x = 2t^2$ $y = \frac{1}{2}t + 1$. Make sure to indicate the direction of the curve.

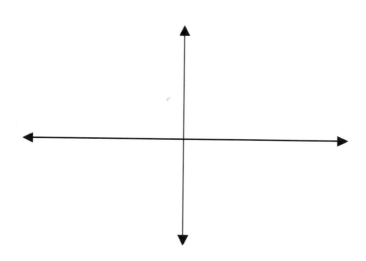

3. Use your graphing utility to graph the following parametric equation: $x = 2t^3$ and $y = t^2$. Is it a function?

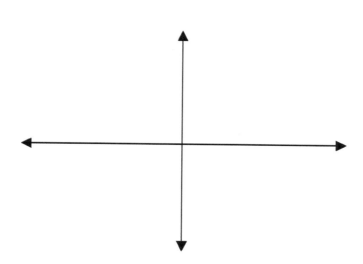

Eliminating the parameter (changing from parametric to rectangular form)

1.) Solve for t in one equation.

2.) Substitute in second equation.

3.) The rectangular equation should result.

4. Eliminate the parameter of $x = 2t$ and $y = \frac{1}{2}t$. Also sketch the graph with the direction indicated.

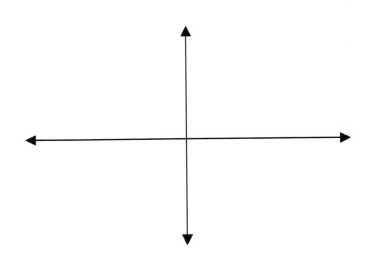

5. Eliminate the parameter of $x = 2\cos\theta$ and $y = 5\sin\theta$. Also sketch the graph with the direction.

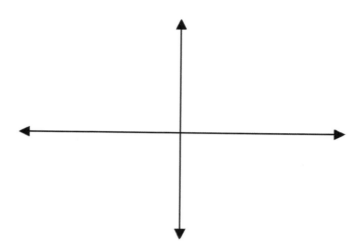

6. Now, find a set of parametric equations to represent the graph given by

$y = 2x^2 + 3$ using the following parameters.

a.) $t = x$

b.) $t = x - 2$

Lecture 8.6 Polar Coordinates

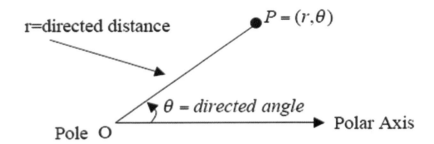

Polar Coordinate system

a two-dimensional coordinate system in which each point on a plane is determined by a distance from a reference point and an angle from a reference direction.

You find a location based upon a direction (θ) and a distance in that direction (r).

1. Plot the following points in the polar coordinate system.

a.) $\left(3, \frac{\pi}{4}\right)$

b.) $\left(4, \frac{-\pi}{6}\right)$

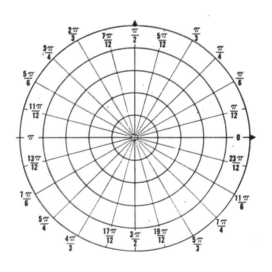

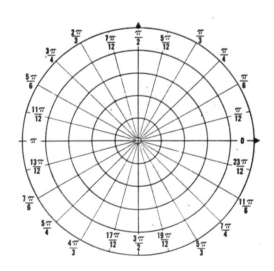

c.) $\left(-2, \dfrac{5\pi}{6}\right)$

d.) $\left(-4, -\dfrac{2\pi}{3}\right)$

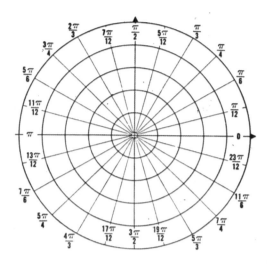

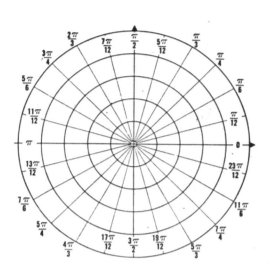

2. Plot the point $\left(1, \dfrac{\pi}{3}\right)$. Then find three other polar representations for the point.

<u>Three other representations:</u>

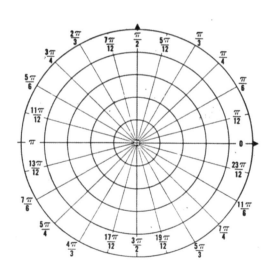

Converting from Polar (r, θ) to Rectangular (x, y) and from Rectangular to Polar.

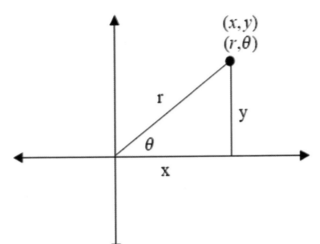

$$\sin \theta = \frac{y}{r} \qquad \cos \theta = \frac{x}{r} \qquad \tan \theta = \frac{y}{x}$$

$$x = r \cos \theta$$

$$y = r \sin \theta$$

$$r = \pm\sqrt{x^2 + y^2}$$

3. Convert the following polar coordinates to rectangular:

a.) $\left(6, \frac{\pi}{6}\right)$

b.) $\left(-4, \frac{5\pi}{4}\right)$

4. Convert the following rectangular coordinates to polar coordinates:

a.) (-2, -2)

b.) (0, -3)

5. Convert the polar equations to rectangular form:

a.) $r = 6$

b.) $\theta = \frac{\pi}{4}$

c.) $r = 2 \cos \theta$

d.) $r = \frac{2}{1 + \sin \theta}$

6. Convert the rectangular equations to polar form:

a.) $x^2 + y^2 - 6x = 0$

b.) $3x - 6y + 2 = 0$

c.) $y^2 = 2x$

d.) $y = x$

Lecture 8.7 Graphs of Polar Equations

Sketch the graph of $r = 2\sin\theta$. Fill in the table and plot the points. Use your graphing calculator to fill in the chart, then plot the points.

Steps to use graphing calculator to fill out table:

1.) Change calculator to Polar Mode and radian.
2.) Enter equation in y=.
3.) Change TBLSET to start at 0 and change in angle equal to 30
4.) View table by pressing 2nd Graph.
5.) Copy the values into the chart below

θ(degrees)	0	30°	60°	90°	120°	150°	180°	210°	270°	330°	360°
radians	0	$\dfrac{\pi}{6}$	$\dfrac{\pi}{3}$	$\dfrac{\pi}{2}$	$\dfrac{2\pi}{3}$	$\dfrac{5\pi}{6}$	π	$\dfrac{7\pi}{6}$	$\dfrac{3\pi}{2}$	$\dfrac{11\pi}{6}$	2π
r											

To graph polar function on calculator:

1.) If in degree mode, change to radian mode and set your *step size* to 0.1.
2.) Press y= and enter the function properly.
3.) Use zooming technique to see the graph.

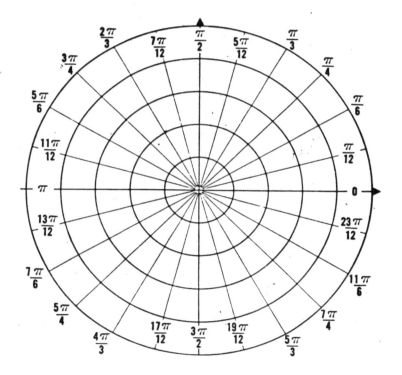

Symmetry tests that can be used on Polar Coordinates

These are true if the given substitution yields an equivalent equation

1.) The line $\theta = \frac{\pi}{2}$: Replace (r, θ) by $(r, \pi - \theta)$ or $(-r, -\theta)$.

2.) The polar axis: Replace (r, θ) by $(r, -\theta)$ or $(-r, \pi - \theta)$.

3.) The pole: Replace (r, θ) by $(r, \pi + \theta)$ or $(-r, \theta)$.

2. Determine the type of symmetry that the following polar graph has:
$r = 2 - 2\sin\theta$. Then, sketch the graph. (you do not have to fill out the entire table if there is symmetry.)

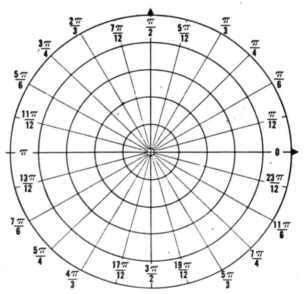

θ(degrees)	0	30°	60°	90°	120°	150°	180°	210°	270°	330°	360°
radians	0	$\frac{\pi}{6}$	$\frac{\pi}{3}$	$\frac{\pi}{2}$	$\frac{2\pi}{3}$	$\frac{5\pi}{6}$	π	$\frac{7\pi}{6}$	$\frac{3\pi}{2}$	$\frac{11\pi}{6}$	2π
r											

3. Sketch the graph of $r = 2$

4. Sketch the graph of $\theta = \frac{\pi}{4}$.

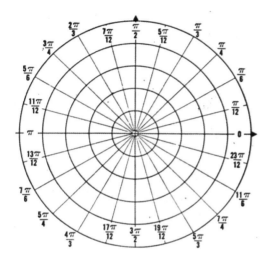

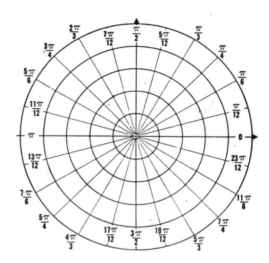

5. Sketch the graph of $r = 3 \sin 3\theta$. Identify the points where $|r|$ is maximum and points where $r = 0$. Use this to help you graph the polar graph.

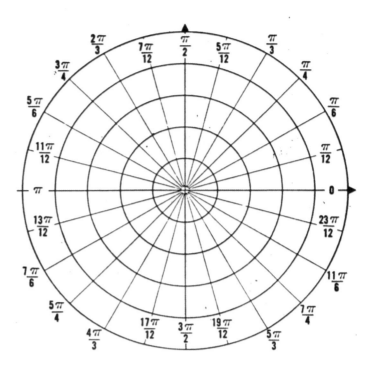

θ(degrees)	0	30°	60°	90°	120°	150°	180°	210°	270°	330°	360°
radians	0	$\dfrac{\pi}{6}$	$\dfrac{\pi}{3}$	$\dfrac{\pi}{2}$	$\dfrac{2\pi}{3}$	$\dfrac{5\pi}{6}$	π	$\dfrac{7\pi}{6}$	$\dfrac{3\pi}{2}$	$\dfrac{11\pi}{6}$	2π
r											

6. Sketch the graph of $r = 1 + 2\cos\theta$.

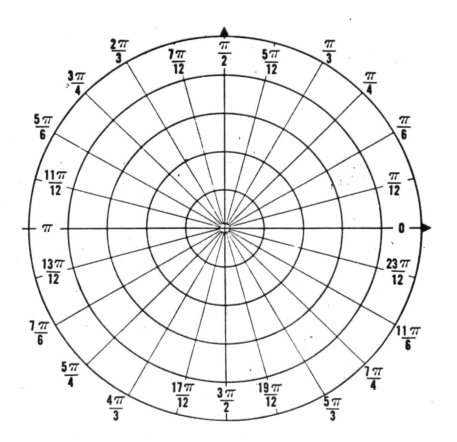

θ(degrees)	0	30°	60°	90°	120°	150°	180°	210°	270°	330°	360°
radians	0	$\dfrac{\pi}{6}$	$\dfrac{\pi}{3}$	$\dfrac{\pi}{2}$	$\dfrac{2\pi}{3}$	$\dfrac{5\pi}{6}$	π	$\dfrac{7\pi}{6}$	$\dfrac{3\pi}{2}$	$\dfrac{11\pi}{6}$	2π
r											

9

Limit and Calculus

Learning Objective

Know how to evaluate Limits and understand the connection between Limits and Derivatives.

Lecture 9.1 Introduction to Limits

Limits

$\lim\limits_{x \to a} f(x)$ => asks us to find the value of f(x) when x is approaching the value of a.

$\lim\limits_{x \to a} f(x) = b$ means that if f(x) becomes arbitrarily close to a unique number b as

x gets closer to a, then the limit of f(x) as x approaches c is b.

1. Use a table to evaluate the following limits with the aid of your calculator. (We must look at the function's value from both sides of the c)

a.) $\lim\limits_{x \to 3} (x - 2)$

x	2.9	2.99	3	3.001	3.01	3.1
f(x)						

b.) $\lim\limits_{x \to -1} \dfrac{x^2 - 1}{x + 1}$

x	-1.1	-1.01	-1	-0.999	-0.99	-0.9
f(x)						

2. Use the graph of the function to evaluate the following limits. (Use your calculator.)

$a.)\ \lim_{x \to 3} \dfrac{x-3}{\sqrt{x+3}-\sqrt{x}}$

$b.)\ \lim_{x \to \pi}(\cos x)$

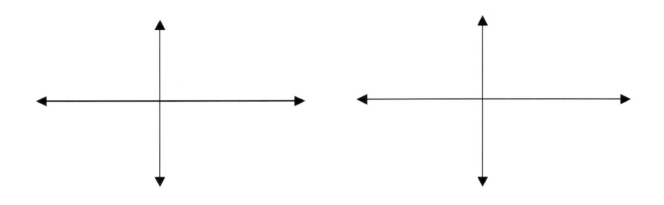

Conditions Under which limits do not exist

1.) $f(x)$ approaches a different number from the right side of c than from the left side of c. → very important in AP.
2.) $f(x)$ increases or decreases without bound as x approaches c.
3.) $f(x)$ oscillates between two fixed values as x approaches c.

3. Do the following limits exist? (Why or why not)

$a.)\ \lim_{x \to 3} \dfrac{|x-3|}{x-3}$

$b.)\ \lim_{x \to 0} \dfrac{2}{x}$

$c.)\ \lim_{x \to 0} \sin \dfrac{1}{x}$

Properties of Limits

1.) $\lim_{x \to c} b = b$ 2.) $\lim_{x \to c} x = c$ 3.) $\lim_{x \to c} x^n = c^n$ 4.) $\lim_{x \to c} \sqrt[n]{x} = \sqrt[n]{c}$

If function is continuous, $f(c)$ is the value of the limit.

Operations with Limits that can be used:

1.) $\lim_{x \to c} bf(x) = b \lim_{x \to c} f(x)$ 2.) $\lim_{x \to c} |f(x) \pm g(x)| = \lim_{x \to c} f(x) \pm \lim_{x \to c} g(x)$

3.) $\lim_{x \to c} [f(x)g(x)] = \left(\lim_{x \to c} f(x) \right) \left(\lim_{x \to c} g(x) \right)$

4.) $\lim_{x \to c} \frac{f(x)}{g(x)} = \frac{\lim_{x \to c} f(x)}{\lim_{x \to c} g(x)}$, $if \lim_{x \to c} g(x) \neq 0$ 5.) $\lim_{x \to c} [f(x)]^n = \left(\lim_{x \to c} f(x)^n \right)$

4. Evaluate the following limits using the properties and operations of limits.

a.) $\lim_{x \to 9} 2\sqrt{x}$ b.) $\lim_{x \to -1} -x^2 - 3x + 7$ c.) $\lim_{x \to 0} \frac{3x+1}{x^2-3x-4}$

d.) $\lim_{x \to 0} (12)$ e.) $\lim_{x \to 0} (2 sinx)$ f.) $\lim_{x \to 3} |x - 3|$

5. Find the limit of $f(x)$ as x approaches 3., where $g(x) = \begin{cases} 7, x \neq 3 \\ -2, x = 3 \end{cases}$

6. Find the limit of $f(x)$ as x approaches 0., where $g(x) = \begin{cases} 3x - 7, x \geq 0 \\ -x^2 - 7, x < 0 \end{cases}$

Lecture 9.2 Techniques for Evaluation Limits

Technique 1.

If p is a polynomial function, then $\lim\limits_{x \to c} p(x) = p(c)$.

If $r(x) = \dfrac{p(x)}{q(x)}$ where p and q are polynomial functions,

then $\lim\limits_{x \to c} r(x) = r(c)$, if $q(c) \neq 0$

1. Evaluate the limits:

$a.)$ $\lim\limits_{x \to 1} 3x^2 + x - 5$

$b.)$ $\lim\limits_{x \to 2} \dfrac{3x^2 - 5x - 1}{x^2 + 1}$

Technique 2.

If direct substitution yields $\dfrac{0}{0}$, then you should use the dividing out technique, where you factor and reduce the fraction before substitution.

2. Evaluate the limits:

$a.)$ $\lim\limits_{x \to 1} \dfrac{x^2 - 1}{x - 1}$

$b.)$ $\lim\limits_{x \to 2} \dfrac{x - 2}{x^3 - 8}$

Technique 3.

Another technique that may need to be used to evaluate limits is called the

rationalizing technique, where you rationalize the numerator when you get $\frac{0}{0}$

3. Evaluate the limit: $\lim\limits_{x \to 0} \frac{\sqrt{x+2}-\sqrt{2}}{x}$

A final technique is using your table feature and your graphing calculator.

4. Find the limit: $\lim\limits_{x \to 0} \frac{1-\cos x}{x}$

One-Sided Limits

$\lim\limits_{x \to C^+} f(x) = L$ is the limit of f(x) as x approaches c from the right of c.

$\lim\limits_{x \to C^-} f(x) = L$ is the limit of f(x) as x approaches c from the left of c.

5. Find the limit of $f(x)$ as x approaches 3.

a.) $f(x) = \dfrac{|x-3|}{(x-3)}$

b.) $f(x) = \begin{cases} x + 6, x \geq 3 \\ x^2, x < 3 \end{cases}$

6. For $f(x) = x^2 - 2x$, find $\lim\limits_{h \to 0} \dfrac{f(3+h) - f(3)}{h}$.

Lecture 9.3 The Tangent Line Problem

Tangent lines can be used to describe the rate at which a curve rises or falls.

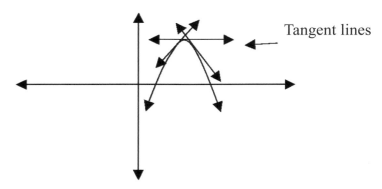

Tangent lines

A tangent line to a curve at a certain point is the line that best approximates the slope of the graph at the point.

1. Approximate the slope of the graph at (-1,1)

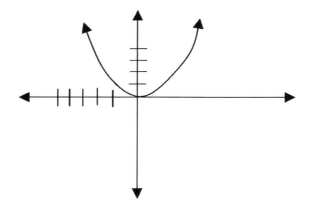

Introduction to Derivative.

The slope of the secant line is $\frac{f(x+h)-f(x)}{h}$. where h can be either small or large.

The slope of the tangent line to the graph of $y = f(x)$ at $(x, f(x))$ is

$m = \lim\limits_{h \to 0} \frac{f(x+h)-f(x)}{h}$.

2. Find the slope of the graph of $y = x^2 at$ (1,1).

3. Find the slope of the graph of $f(x)=x^2 - 2$ at x.

The Derivative of a function f(x) is $f(x) = \lim\limits_{h \to 0} \frac{f(x+h)-f(x)}{h}$. provided the limit exist.

Various notations of the derivate of y:

$$\frac{dy}{dx} = y' = \frac{d}{dx}[f(x)] = f'(x)$$

4. Find the derivative of the following function: $f(x) = x^2 + x - 3$

5. Find the derivative of the following function: $f(x) = \frac{1}{x}$. Then, find the slope of

the graph f at the points (1, 1) and $(-2, \frac{-1}{2})$

Lecture 9.4 Limits at Infinity and Limits of Sequences

1. Draw the graph of $f(x) = \dfrac{x}{x-2}$. Find the $\displaystyle\lim_{x\to\infty} f(x)$ and $\displaystyle\lim_{x\to-\infty} f(x)$.

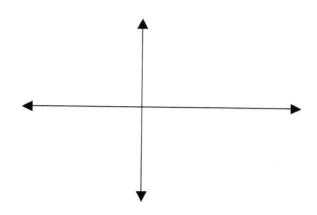

Infinite Limits.

If r is a positive real number, then $\displaystyle\lim_{x\to\infty}\frac{1}{x^r} = 0$. Furthermore, if x^r is defined for

$x<0$, then $\displaystyle\lim_{x\to\infty}\frac{1}{x^r} = 0$.

Evaluate the following limits:

2. $\displaystyle\lim_{x\to\infty} \frac{2x^4 - 5}{x^4}$

3. $\displaystyle\lim_{x\to\infty} \frac{3x-1}{x^2}$

4. $\displaystyle\lim_{x\to\infty} \frac{x^2-1}{5x^2}$

5. $\displaystyle\lim_{x\to\infty} \frac{2x^3-1}{7x^2-x}$

Infinite Limit

For $f(x) = \frac{N(x)}{D(x)}$, where $N(x) = a_n x^n + \cdots + a_0$ and $D(x) = b_m x^m + \cdots + b_0$

$$\lim_{x \to \pm\infty} f(x) = \begin{cases} 0, n < m \\ \dfrac{a_n}{b_n}, n = m \end{cases}$$

6. Find the limit of the following sequences:

a.) $\lim\limits_{n \to \infty} \dfrac{3n+5}{2n^2+1}$

b.) $\lim\limits_{n \to \infty} \dfrac{5n^2}{2n^2+1}$

c.) $\lim\limits_{n \to \infty} \left[\dfrac{2}{n^4} \left(\dfrac{n^2(n+1)^2}{3} \right) \right]$

WorkSheet

PreCalculus Worksheet.

It is prepared so that you can learn the important parts that you studied in PreCalculus Note once more thoroughly.

PreCalculus Section 1.1 Work Sheet

1. $f = \{(2,3), (-3,4), (4,3)\}$

   a.)   What is the domain of f? _____

   b.)   What is the range of f? _____

   c.)   Is f a function? _____

2. Which of the following equations represents y as a function of x? (Answer yes or no)

   a.) $x^2 - 2y + 3 = 0$

   _____

   b.) $2x + y^2 - 5 = 0$

   _____

3. $f(x) = x^2 - 3x + 2$, find the following

   a.) $f(2) =$ _____

   b.) $f(a) =$ _____

   c.) $f(x + h) =$ _____

4. Find the domain of the following functions:

   a.) $f(x) = x^3 + 3x - 2$ domain= _____

   b.) $f(x) = \frac{-2}{x+3}$ domain= _____

   c.) $f(x) = \sqrt[4]{3 - x}$ domain= _____

5. $f(x) = x^2 + 2x - 1$, $find \frac{f(x+h)-f(x)}{h} =$ _____

6. $f(x) = x^3 - 5x + 2$,   find $\frac{f(x+h)-f(5)}{x+h-5} =$ _____

PreCalculus Section 1.2 Work Sheet

Find the domain and range of the following functions:

1. $f(x) = x^3 - 4x + 3$.   Domain= _____   Range=_____

2. $h(x) = -|x - 2|$ Domain=_____   Range=_____

Determine whether y is a function of x.

3. $2x - y^2 = 1$ _____

4. $y = \frac{1}{5}x^5$ _____

Determine the intervals over which the function is increasing, decreasing, or constant and determine whether the function is even, odd, or neither.

5. $f(x) = x^2 - 4x$

Increasing_____   Decreasing_____ Constant _____

Is it even or odd?   _____

6. $f(x) = -x^6 - 3x^4$

Increasing_____Decreasing_____ Constant _____

Is it even or odd? _____

7. Graph the piecewise-defined function

a.) $f(x) = \begin{cases} x^2 + 4, x \le 1 \\ -x^2 + 4x + 3, x > 1 \end{cases}$

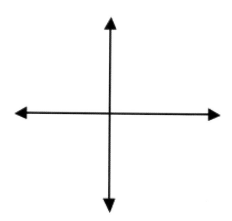

Is the function continuous? _____

b.) $f(x) = \begin{cases} x^2 - 4, x \le 2 \\ \ln(x - 1), x > 2 \end{cases}$

Is the function continuous? _____

PreCalculus Section 1.3 Work Sheet

1. Sketch the graph of the three functions:

a.) $f(x) = \sqrt{x}$ b.) $f(x) = \frac{1}{2}\sqrt{x}$ c.) $f(x) = -\frac{1}{2}\sqrt{x-3}$

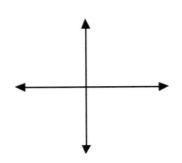 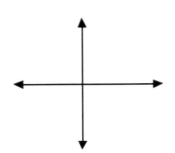 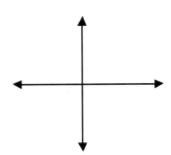

2. Write the equations for the following common functions: (asumme 1 interval = 1 unit.)

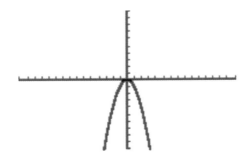 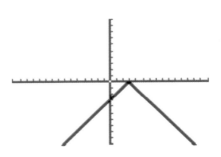

a.) _____   b.) _____

3. $g(x) = 2(x - 3)^2$

   a.) Identify the common function f that is related to g.

   a.) _____

   b.) Describe the sequence of transformations from f to g.

   b.) _____

   c.) Sketch the graph of g

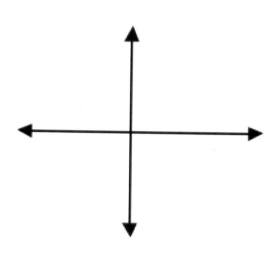

4. $h(x) = -(-x + 1)^2 - 3$

   a.) Identify the common function f that is related to h.

   a.) _____

   b.) Describe the sequence of transformations from f to h.

   b.)_____

   c.) Sketch the graph of h

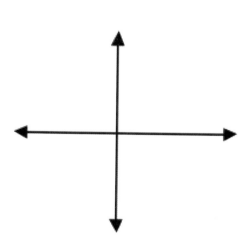

PreCalculus Section 1.4 Work Sheet

1. $f(x) = 3x - 5$ and $g(x) = 2 - x$. Find the following:

a.) $(f + g)(x) =$ _____

b.) $(f - g)(x) =$ _____

c.) $(fg)(x) =$ _____

d.) $\left(\frac{f}{g}\right)(x) =$ _____

e.)   what is the domain of $\frac{f}{g}$?   _____

f.) $(fg)(2) =$ _____

g.) $(f - g)(-5) =$ _____

2. $f(x) = |x + 1|$, and $g(x) = x - 4$, find the following:

a.) $f \circ g =$ _____

b.) $g \circ f =$ _____

3. Find two functions f and g such that $(f \circ g)(x) = h(x)$.

a.) $h(x) = \frac{4}{(5x+2)^2}$ $f(x) =$ _____

                                            $g(x) =$ _____

1. Determine whether or not the function is one-to-one. (yes or no)

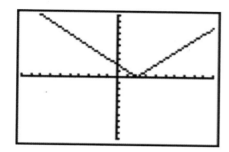

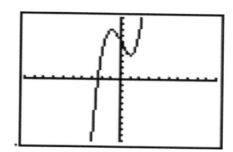

a.) _____

b.) _____

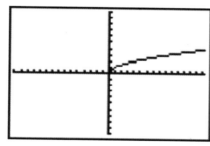

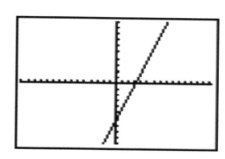

c.) _____

d.) _____

2. Find the inverse of $f(x) = \frac{2x-1}{3}$ algebraically. Also Verify your results showing that $f(f^{-1}(x)) = x$ and $f^{-1}(f(x)) = x$.

Verification

Inverse=_____

3. Determine algebraically whether the function is one-to-one.

a.) $f(x) = \sqrt{x-3}$

Is $f(x)$ one-to-one? _____

4. Are $f(x) = \sqrt[3]{2x-10}$ and $g(x) = \frac{x^3+10}{2}$ inverses of each other?

_____

5. Sketch the graph of the inverse functions on the following graphs:

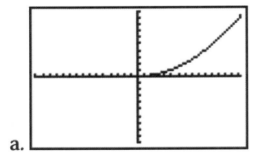

a.

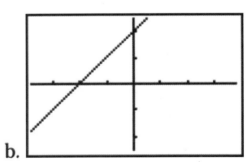

b.

1. What is the degree of the following functions?

a.) $h(x) = 2 - 4x^4 + x^3$ 　　　　　　　　Degree=_____

b.) $y(x) = -3x + 2x^3$ 　　　　　　　　　　Degree=_____

2. Sketch the following functions: (identify vertex, intercepts, and axis of symmetry)

a.) $f(x) = 2x^2 - x + 1$

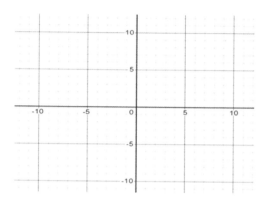

b.) $g(x) = -3x^2 + 9x + 1$

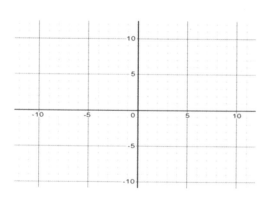

3. Find the quadratic function that has the indicated vertex and whose graph passes through the given point. Vertex: (4,-1) Point (2,3)

4. Find the quadratic function that has the indicated vertex and whose graph passes through the given point. Vertex: (2,3) Point (0,2)

5. Determine the coordinates of the vertex and the equation of the axis of symmetry of $y = 3x^2 + 2x - 5$. Does the quadratic function have a minimum or maximum value? If so, what is it?

PreCalculus Section 2.2 Work Sheet

1. Match the polynomial function with its graph:

   a.) $y = -2x + 3$

   b.) $y = -2x^2 + 2x + 1$

   c.) $y = x^4 + 3x + 1$

   d.) $y = \frac{1}{5}x^5 - 2x^3 + \frac{9}{5}x$

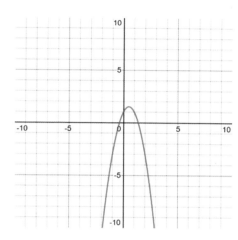

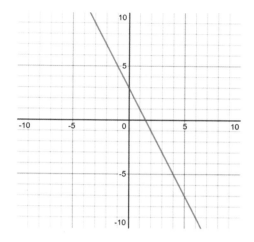

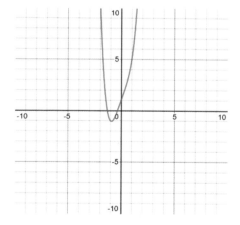

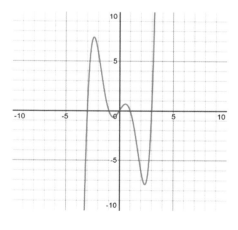

2. Use the Leading Coefficient Test to determine the right-hand and left-hand behavior of the graph of the polynomial function.

a.) $f(x) = \frac{2}{3}x^3 - 3x$ _____

b.) $h(x) = 2 - 3x^6$ _____

3. Find all the real zeros of the polynomial function.

a.) $f(x) = 81 - x^2$

_____

b.) $(x) = x^4 - 6x^3 - 7x^2$

_____

4. Graph the following function: $f(x) = -4x^3 + 4x^2 + 15x$

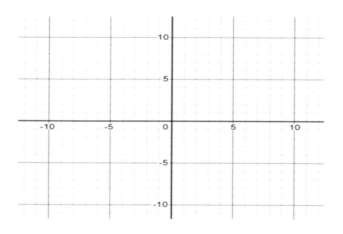

5. Find a polynomial function that has the given zeros.

a.) -5, -1, 0, 1, 2

PreCalculus Section 2.3 Work Sheet

1. Divide $x^3 - 9$ by $x^2 + 1$ using long division.

2. Divide $5x^3 + 6x + 8$ by $x + 2$ using synthetic division.

3. What is the remainder when $3x^3 + 2x^2 - 5x - 8$ is divided by $x + 2$?

4. If $3 + 2i$, 2, and $2 - 3i$ are all zeros of $P(x) = 3x^5 - 36x^4 + 2x^3 - 8x^2 + 9x - 338$, what are the other zeros?

5. Use synthetic division to show that x is a solution of the third-degree polynomial equation and use the result to factor the polynomial completely. List all the real zeros of the function. $x^3 - 28x - 48 = 0$, $x = -4$ is a solution.

Complete Factorization_____

All real Zeros_____

6. Use the Rational Zero Test to list the possible rational roots and then find all of the rational roots. $f(x) = x^3 - 4x^2 - 4x + 16$

Possible Rational Roots_____

Rational Roots_____

7. Find all the real solutions of $x^4 - 13x^2 - 12x = 0$

Solutions_____

PreCalculus Section 2.4 Work Sheet

1. Solve for a and b; $(a + 6) + 2bi = 4 - 5i$ a=_____   b=_____

2. Write in standard form: $-\sqrt{-75} + 3$ _____

3. Simplify and write in standard form:

a.) $(13 - 2i) + (-3 + 6i) =$ _____

b.) $(6 - 2i)(2 - 3i) =$ _____

c.) $\left(3 + \sqrt{-5}\right)\left(7 - \sqrt{-10}\right) =$ _____

4. Find the product of the number and its conjugate:

   a.) $(7 - 12i)$ _____

5. Simplify and write in standard form:

a.) $\left(\dfrac{5}{1-i}\right)$ _____

b.) $\left(\dfrac{8-7i}{1-2i}\right)$ _____

6. Simplify the complex number and write in standard form:

a.) $i^2 + i^{23} - 7i$ _____

b.) $\dfrac{3}{(2i)^3}$ _____

7. Plot and label the complex numbers in the complex plane:

   a.) $2 + 3i$ b.) $-3i + 5$ c.) $5i$ d.) -3   e.) $5 - 3i$

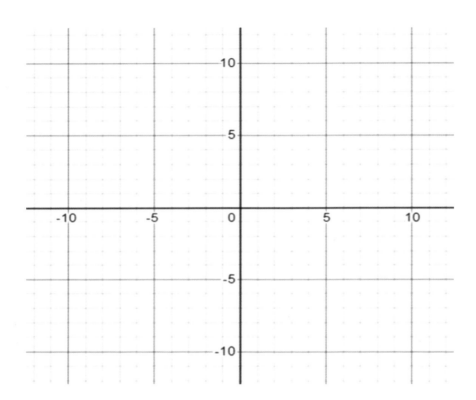

PreCalculus Section 2.5 Work Sheet

1. Find all the zeros of the function and write the polynomial as a product of linear factors.

a.) $g(x) = x^2 + 10x + 23$ Zeros_____

Factors_____

or_____

b.) $h(x) = x^3 - 3x^2 + 4x - 2$

Zeros_____

Factors _____

2. Find a polynomial function with integer coefficients that has the given zeros: 4, 3i, -3i

Function_____

3. Given: $f(x) = x^4 + 6x^2 - 27$

a.) write $f(x)$ as the product of factors irreducible over the rationals:

_____

b.) write $f(x)$ as the product of linear and quadratic factors that are irreducible over the reals:

_____

c.) write $f(x)$ in completely factored form:

_____

4. Find all the zeros of $f(x) = 2x^4 - x^3 + 7x^2 - 4x - 4$ given 2i is a root.

Zeros_____

Complete Factorization _____

Or _____

PreCalculus Section 2.6 Work Sheet

1. Find the domain and the asymptotes of the following functions:

a.) $f(x) = \frac{x+5}{x^2+2x-3}$ Domain:_____ Horizontal Asymptote:_____

Vertical Asymptote:_____

b.) $f(x) = \frac{x^2-8}{2x-9}$ Domain:_____ Vertical Asymptote: _____

Slant Asymptote: _____

2. Given: $f(x) = \frac{x}{2x-5}$ find the following and then graph:

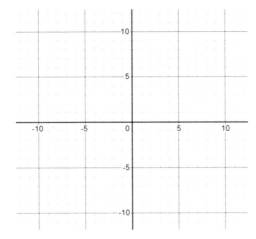

Domain: _____

Vertical Asymptote: _____

Horizontal Asymptote: _____

3. Given: $f(x) = \dfrac{x^2}{x^3 - 8}$

Domain: _____

Vertical Asymptote: _____

Horizontal Asymptote: _____

Precalculus Section 2.7 Work Sheet

1. Sketch the graph of $f(x) = \frac{1-2x}{x}$

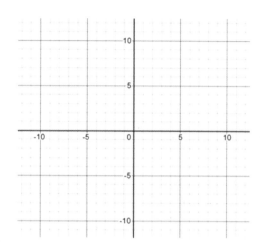

x intercepts: _____

y intercepts:_____

vertical asymptotes: _____

horizontal asymptotes: _____

slant asymptotes: _____

Symmetry: _____

Additional Points

x					
f(x)					

2. Sketch the graph of $g(x) = \frac{x^3}{2x^2 - 8}$

x intercepts: _____

y intercepts:_____

vertical asymptotes: _____

horizontal asymptotes: _____

slant asymptotes: _____

Symmetry: _____

Additional Points

x					
f(x)					

Precalculus Section 3.1 Work Sheet

1. Use your calculator to evaluate $e^{2.58}$ to three decimal places: _____

2. Graph and label the following functions on the same graph: (label intercepts and asymptotes)

a.) $f(x) = \left(\frac{5}{2}\right)^x$

b.) $g(x) = \left(\frac{5}{2}\right)^{-x}$

c.) $h(x) = \left(\frac{5}{2}\right)^{x+2}$

d.) $i(x) = -\left(\frac{5}{2}\right)^{x+2} - 1$

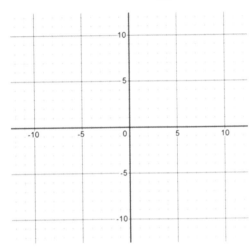

3. Use a graphing calculator to construct a table of values, then sketch the graph of

$f(x) = e^{-x}$

x	f(x)

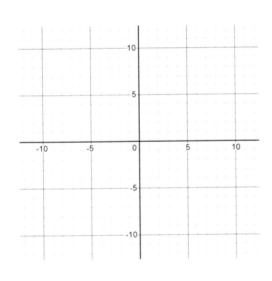

4. Complete the table to determine the balance A for $12,000 invested at a rate $r = 6\%$ compounded continuously for t years.

t	1	10	20	30	40	50
A						

5. The population of a town increases according to the model $P(t) = 2500e^{.0293t}$ where t is the time in years, with t=0 corresponding to 1990.

Find the population in 1992, 1995, and 1998.

1992=_____   1995=_____   1998=_____

PreCalculus Section 3.2 Work Sheet

1. Write the logarithmic equation in exponential form

    a.) $\log_3 27 = 3$ _____

    b.) $\log_{16} 2 = \frac{1}{4}$ _____

2. Write the exponential equation in logarithmic form.

    a.) $5^3 = 125$ _____

    b.) $e^x = 3$ _____

3. Evaluate the expression without a calculator.

    a.) $\log_{27} 9$ _____

    b.) $\log_2 \left(\frac{1}{8}\right)$ _____

4. Solve the equation for x.

    a.) $\log_7 7 = x$ _____

    b.) $\log_3 3^{-1} = x$ _____

5. Use a calculator to evaluate the logarithm. Round to three decimal places.

    a.) $\log_{10} 65$ _____

    b.) $-8.5 \ln 14$ _____

6. Find the domain, vertical asymptote, x intercept and sketch its graph by hand. Verify using your graphing calculator.

a.) $g(x) = \log_{10} x$

Domain : _____

Vertical asymptote : _____

x-intercept : _____

b.) $f(x) = -\log_{10}(x + 3)$

Domain : _____

Vertical asymptote : _____

x-intercept : _____

PreCalculus Section 3.3 Work Sheet

1. Evaluate the logarithm using the change-of-base formula. Round your result to three decimal places.

   a.) $\log_3 42$ _____

   b.) $\log_{\frac{1}{6}} 36$ _____

   c.) $\log_{\frac{1}{3}}(615)$ _____

2. Rewrite the logarithm as a multiple of (a) a common logarithm and (b) a natural logarithm.

   $\log_x \frac{3}{4}$ a.)_____b.) _____

3. Use the properties of logarithms to write the expression as a sum, difference, and / or constant multiple of logarithms. (Assume all variables are positive)

   a.) $\log_{10}(10a)$

   b.) $\log_6 b^{-3}$

   c.) $\ln\left(\frac{x^3-1}{2x^3}\right), x > 1$

   d.) $\ln\sqrt{x^2(x+2)}$

4. Write the expression as the logarithm of a single quantity.

a.) $\ln a + 2\ln b$

b.) $\frac{7}{2}\log_e(c - 4)$

c.) $2\ln 6 + 5\ln d$

d.) $2[\ln x - \ln(x + 1) - \ln(x - 1)]$

5. Find the exact value of the logarithm without using a calculator, if possible. If not possible, state why.

a.) $\log_5 \sqrt[3]{5}$

b.) $\log_4 4 + \log_4 16$

c.) $\log_4(-64)$

Precalculus Section 3.4 Work Sheet

1. Use a graphing utility to graph f and g in the same viewing window. Approximate the point of intersection of the graphs. Then solve the equation $f(x) = g(x)$ algebraically.

a.) $f(x) = 9$ $g(x) = 27^x$ Approximation _____

                                        Algebraic Solution: _____

b.) $f(x) = 3 \log_5 x$ $g(x) = 6$ Approximation _____

                                        Algebraic Solution: _____

2. Solve for x.

a.) $3^x = 81$ _____

b.) $\left(\frac{3}{4}\right)^x = \frac{27}{64}$ _____

c.) $\ln(2x + 5) = 7$ _____

3. Simplify the expression

a.) $\ln e^{3x-1}$ _____

b.) e^{3lnx} _____

4. Solve the exponential equation algebraically. Round your result to three decimal places. Use your graphing calculator to verify your results.

a.) $6^{5x} = 3000$

_____

b.) $1000e^{-4x} = 75$

_____

c.) $\dfrac{525}{1+e^{-x}} = \dfrac{275}{1}$

_____

5. Use your graphing calculator to find an approximate solution to three decimal places.

a.) $4^{\frac{-x}{2}} = 0.10$ _____

b.) $\dfrac{119}{e^{6x}-14} = 7$ _____

6. Solve the logarithmic equation algebraically. Round your answer to three decimal places.

a.) $\ln 4x = 1$

_____

PreCalculus Section 3.5 Work Sheet

1. Sketch a scatter plot of the data, decide whether the data could be modeled by a linear, exponential, or a logarithmic model.

   a.) (1,11), (1.5,9.6), (2,8.2), (4,4.5), (6,2.5), (8,1.4)

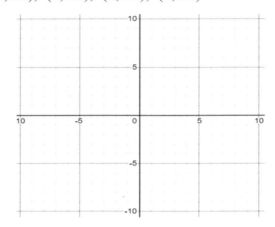

2. Complete the table for the time t necessary for P dollars to triple if interest is compounded annually at rate r.

r	2%	4%	6%	8%	10%	12%
t						

3. The population P of a city is $P = 240{,}360e^{0.012t}$, where t=0 represents 2000. According to this model, when will the population reach 275,000?

PreCalculus Section 4.1 Work Sheet

1. Convert each degree measure to radian.

a.) $150°$

b.) $-225°$

2. Convert each radian measure to degree measure.

a.) $\frac{5\pi}{3}$

b.) $\frac{-12\pi}{6}$

3. Determine the quadrant in which the angle lies.

a.) $\frac{55\pi}{3}$

b.) $\frac{2.35}{2\pi}$

c.) $\frac{-5\pi}{6}$

d.) $\frac{-35\pi}{4}$

4. Find the complement and supplement of $\frac{2\pi}{7}$.

5. Determine the length of the arc of a circle of radius 5.25ft intercepted by a central angle having a measure of 57^0.

PreCalculus Section 4.2 Work Sheet

1. Give the exact value (if defined) of the six trigonometric functions of $\theta = \frac{-5\pi}{6}$.

2. Find the exact value of the $\tan \frac{17\pi}{6}$.

3. Use the function value $\tan t = 3$ to find the values of each of the following:

a.) $\tan(-t) =$ b.) $\cot(t) =$

4. Use a calculator to evaluate the following: (Round to four decimal places)

a.) $\csc(7.89) =$ b.) $\tan(-120.4°) =$

1. Given: $\cos\theta = -\frac{2}{7}$, $and\ \theta$ is in quadrant III. Find the other five trigonometric functions by drawing a triangle.

2. Suppose $\cot\theta = 2\ and\ \theta$ is in the third quadrant. Find the $\csc\theta$ by using a trigonometric identity.

3. Using a calculator, find the following to four decimal places:

a.) $\csc 32.7°$

b.) $\tan 2.45$

$c.$) $\sec 132°$

d.) $\cos -3.53$

4. A 23-foot ladder, leaning against the side of a house, reaches 12 feet up from the bottom corner of the house. What angle does the ladder make with the ground?

PreCalculus Section 4.4 Work Sheet

1. State the quadrant in which θ lies.

a.) $\sec\theta > 0$ and $\csc\theta < 0$

b.) $\cot\theta < 0$ and $\cos\theta < 0$

2. Find the values of the remaining five trigonometric functions if $\sec x = \frac{13}{-5}$ and $\cot x < 0$.

3. Find the reference angle for each of the following:

a.) $\theta = 356°$ Reference angle=_____

b.) $\theta = -453°$ Reference angle=_____

c.) $\theta = \frac{15\pi}{3}$ Reference angle=_____

d.) $\theta = \frac{-7\pi}{3}$ Reference angle=_____

4. Using a calculator, evaluate each of the following to four decimal places.

a.) $\tan 572.7° =$ b.) $\sec 160° =$

5. Using a calculator, find two values of θ , where $0° \leq \theta < 360°$. (Round values to one decimal place.)

a.) $\sin\theta = .4226$ b.) $\cos\theta = -.6018$

1. Find the amplitude, period, the phase shift and vertical displacement (if they exist).

a.) $y = 3\cos\frac{x}{2} + 2$

b.) $y = \frac{1}{2}\sin(3x - \pi)$

c.) $y = -5\sin(2x - \frac{\pi}{2})$

2. Describe how the graph of the first equation can be obtained from the graph of the second equation.

a.) $y = \sin(2x - 2\pi)$

$y = \sin 2x$

b.) $y = -3 + 4\cos(x - \frac{\pi}{4})$

$y = 2\cos(x - \frac{\pi}{4})$

3. Graph each equation, showing one complete cycle of the curve.

a.) $y = 3\cos 2x$

b.) $y = 2\sin 2(x + \frac{\pi}{4}) - 4$

1. Graph each function, showing one complete cycle.

a.) $y = \tan\left(\frac{1}{2}x\right) + 2$

b.) $y = -3 + \frac{1}{2}\sec\left(2\left(x - \frac{\pi}{2}\right)\right)$

c.) $y = 3\csc(2\pi x) - 5$

PreCalculus Section 4.7 Work Sheet

1. Find the exact values of each expression, where $0° \leq \theta < 360°$:

a.) $arc \cos(\frac{-\sqrt{2}}{2}) =$

b.) $arc \sin\left(\frac{-\sqrt{3}}{2}\right) =$

c.) $\cos(arc \tan(-1)) =$

d.) $\sin(arc \tan(\sqrt{3})) =$

e.) $arc \cos\left(\sin\left(\frac{5\pi}{4}\right)\right) =$

f.) $\cos(arc \sin\left(\frac{-\sqrt{3}}{2}\right)) =$

2. Write the following as an algebraic expression in terms of x:

a.) $\cot(arc \tan x) =$

b.) $\sin(arc \cos x) =$

c.) $\sec(arc \csc \frac{2}{x}) =$

PreCalculus Section 5.1 Work Sheet

1. Use the values of $\csc\theta = \frac{-13}{12}$ and $\cos\theta > 0$ to find the values of all six trigonometric functions.

2. If $\sec\theta = -\frac{5}{4}$ and $\csc\theta > 0$, then $\cot\theta =$

3. Simplify each of the following:

a.) $2\sec^2\theta - 2\tan^2\theta =$

b.) $\sec x - \sin x \tan x =$

c.) $\frac{\csc^2 x - 1}{\csc x - 1} =$

d.) $\sec^3 x - \sec^2 x - \sec x + 1 =$

e.) $2\sin^2 x - 18\cos^2 x =$

f.) $6\sin^2 x + 3\sin x - 3 =$

1. Simplify each of the following:

a.) $\csc x \sin x =$

b.) $\sin x + \cot x \cos x =$

c.) $3\tan^3 x - 192 \tan x$

d.) $\dfrac{3\sin x + 6}{4 - \sin^2 x} + \dfrac{4}{\sin x - 2}$

2. Verify each of the following identities:

a.) $\dfrac{\csc \theta}{\sec \theta} = \cot \theta$

b.) $\cos x + \sin x \tan x = \sec x$

c.) $\cos x(\sec x - \cos x) = \sin^2 x$

d.) $\dfrac{\cot^2 x}{1 + \csc x} = \csc x - 1$

1. Solve $\tan^2 x = \tan x + 2$ in the interval: $[0, 2\pi]$.

_____

2. Given: $2\cos x - 1 = 0$

a.) Find the solutions in the interval: $[0, 2\pi)$.

_____

b.) Write the equation(s) for all the possible solutions:

_____

3. Solve $\cos x + \sqrt{2} = -\cos x$ in the interval: $[0, 2\pi)$.

_____

4. Solve $2\sin^2 x = 2 + \cos x$ in the interval: $[0, 2\pi)$.

_____

5. Solve $\sin 2x = \frac{-\sqrt{3}}{2}$ in the interval: $[0, 2\pi)$.

_____

PreCalculus Section 5.4 Work Sheet

1. a.) Find the exact value of $\cos\frac{\pi}{12}$.

   b.) Find the exact value of $\sin 75°$.

   c.) Find the exact value of $\cot 375°$

2. Simplify:

   $a.)$ $\sin\theta\cos 2\theta + \cos\theta\sin 2\theta =$

   $b.)$ $\cos 25°\cos 15° - \sin 25°\sin 15° =$

3. If $\tan A = \dfrac{5}{12}$ and $\sin B = \dfrac{3}{5}$, where A and B are acute angles, find the value of $\cos(A+B)$.

4. Find the exact value of $\tan 75°$.

5. Let $\cos\alpha = \dfrac{4}{5}$ and $\sin\beta = \dfrac{5}{13}$ where $\dfrac{3\pi}{2} < \alpha < 2\pi$ and $\dfrac{\pi}{2} < \beta < \pi$. Find $\sin(\alpha - \beta)$.

PreCalculus Section 5.5 Work Sheet

1. Express $\sin 4x$ in terms of $\sin x$ and $\cos x$.

2. Given that $\pi < \theta < \frac{\pi}{2}$ and $\cos \theta = \frac{-15}{17}$, find the value for each of the following:

   a.) $\sin 2\theta =$

   b.) $\cos 2\theta =$

   c.) $\tan 2\theta =$

   d.) What quadrant is 2θ in?

3. If $\sin \theta = a$, find the value of $\sin 2\theta$ in terms of a.

4. Using the half-angle formula, find the exact value of $\tan 75°$.

5. Solve $\sin x \sin 40° - \cos x \cos 40° = \frac{1}{2}$ for all value of x such that $0° \leq x \leq 360°$.

6. For what values of x between 0 and 2π is $\sin x < \cos x$?

7. If $\cos 20°=b$, find the value of $\cos 40°$ in terms of b.

1. Given $\sin \theta = \frac{1}{5}$. Find the $\cos \theta$ if θ is in the 2nd quadrant.

Simplify the following:

2. $\sin x \tan x \cos x$

3. $\csc A \tan A \cos A$

4. $\frac{\sin X}{\cos X} + \frac{\cos X}{\sin X}$

5. $\frac{\sin^2 \theta - \cos^2 \theta}{\sin \theta - \cos \theta}$

6. $\cot \theta + \frac{1 - 2 \cos^2 \theta}{\sin \theta \cos \theta}$

7. Verify that $\cos^2 \theta - \sin^2 \theta = 1 - 2 \sin^2 \theta$.

8. Verify that $\frac{1}{1 - \cos x} + \frac{1}{1 + \cos x} = 2 + 2 \cot^2 x$

9. Find the solutions in the interval $[0, 2\pi)$ of $\cos^2 x + \cos x - 2 = 0$.

10. Find the solutions in the interval $[0, 2\pi)$ of $2 \cos x - 1 = 0$.

11. Find the exact value of the $\cos 195°$.

12. Find the exact value of the $\tan 15°$.

13. Suppose $\sin \theta = \frac{4}{5}$ and $\frac{\pi}{2} < \theta < \pi$, find the $\sin 2\theta, \ \cos 2\theta$., and $\tan 2\theta$.

14. Find the solutions on $[0,2\pi)$ of $\sin^2 x + 2 \cos x = 2$.

15. Find the exact value of the trigonometric functions given that the $\sin u = \frac{3}{4}, \quad \cos v = \frac{-5}{13}$ and u and v are in Quadrant II.

a.) $\sin(u + v) =$ 　　　　　　　　　　　 b.) $\cos(u - v) =$

c.) $\tan(2u) =$

PreCalculus Section 6.1 Work Sheet

1. Given the following values, determine how many triangles can be formed.

a.) a=15.2, b=8.5, and ∠B=42°

_____

b.) c=28, b=11, and ∠B=40°

_____

c.) b=8, a=7, and ∠B=30°

_____

2. Given triangle ABC with ∠A=39°, ∠B=106°, and c=78. Find a.

_____

3. Find all six parts of triangle ABC, given ∠A=48°, ∠C=57°, and b=47.

_____

4. A cellular phone signal tower sits on the ground. Two 88-foot guy wires are positioned on opposite sides of the tower. The angle of elevation each wire makes with the ground is 21°. How far apart are the ends of the two guy wires?

_____

PreCalculus Section 6.2 Work Sheet

1. Find the number of degrees in the other two angles of $\triangle ABC$ if $c = 75\sqrt{2}$, $b = 150$, and $\angle C = 30°$

    _____

2. In triangle ABC, a=9, b=13, and $\angle C=82°$. Find the length of c.

    _____

3. In triangle ABC, a=13.2, b=18.5, and c=26.2. Find the largest angle.

    _____

4. A tunnel is to be dug from point A to point B. The distance from a third point C to A is 3.65 miles and from point C to B is 2.74 miles, and $\angle ACB=49.2°$. How long will the tunnel be in length?

    _____

5. A surveyor finds the edges of a triangular lot to measure 5.3m, 10.5m, and 14m. Find the area of the lot.

    _____

6. Find the area of $\triangle ABC$ if $a = 180$ inches, $b = 150$ inches, and $\angle C = 30°$

    _____

PreCalculus Section 6.3 Work Sheet

1. Given the vectors a, b, c, and d sketch the following:

   a.)   2a + c

   b.) 3d – 2b

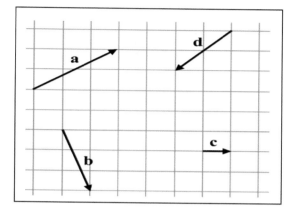

2. Given the vectors a, b, c, and d find the component of each vector (assume 1 unit = 1):

   a.) 2b

   b.) -2c+d

   b.) 2a+c

3. A vector v has initial point (4,10) and terminal point (-4, -5). Find the following:

   a.) Component Form: _____

   b.) $\|v\|$ _____

   c.) Direction of the vector in component form: _____

   d.) A unit vector u in the direction of v.   _____

4. Let u = (4,5) and v = (8,-5). Find each of the following:

   a.) u – v =

   b.) 2u + 3v =

   c.) Write u as a linear combination of i and j.   _____

PreCalculus Section 7.1 Work Sheet

1. Simplify: $a.)$ $\frac{5!}{4!3!} =$

   a._____

   $b.)$ $\frac{2(n+2)!}{3n!} =$

   b._____

2. Write the first four terms of the sequence whose n^{th} term is $a_n = \frac{2n}{n^2+1}$.

   _____

3. Find a formula for the nth term of the sequence:

   -7, 2, 11, 20, 29, 38,…

   _____

4. Write the series in expanded form and evaluate:

   $\sum_{i=2}^{6}(3i - 1) =$

   Sum=_____

5. Use sigma notation to write the sum:

   a.) $1 + \frac{1}{2} + \frac{1}{4} + \frac{1}{8} + \frac{1}{16}$

   Sigma Notation:_____

   b.) $2 + \frac{2}{2} + \frac{2}{5} + \frac{2}{10} + \frac{2}{17}$

   Sigma Notation:_____

6. Write the first four terms of the sequence defined recursively:

   $a_1 = 3$ and $a_{k+1} = 2a_k - 1$

   _____

PreCalculus Section 7.2 Work Sheet

1. Determine whether the sequence is arithmetic. If it is, find the common difference.

a.) $3, 9, 15, 21, 27, \ldots.$ _____

b.) $1^2, 2^2, 3^2, 4^2, 5^2, \ldots$ _____

c.) $\frac{1}{2}, \frac{1}{4}, \frac{1}{6}, \frac{1}{8}, \ldots$ _____

2. Write the first four terms of the arithmetic sequence:

a.) $a_1 = 13$ and $a_{b+1} = a_b + 5$ _____

3. Write the first four terms of the arithmetic sequence with $a_1 = 13$ and $d = -2$.

_____

4. Find the 12th term of the arithmetic sequence with $a_1 = -23$ and d=-4.

_____

5. Find the sum of the first 25 terms of the arithmetic sequence: $15, 25, 35, 45, 55, \ldots$

_____

1. Write the first four terms of the geometric sequence with $a_1 = 6$ and $r = \frac{3}{2}$.

    _____

2. Find the seventh term of the geometric sequence: $\frac{5}{64}, -\frac{5}{16}, \frac{5}{4}, \ldots$

    _____

3. Find the sum: $\sum_{k=2}^{7} 5^{k-2}$

    _____

4. Find the sum of the infinite geometric series: $8 - 4 + 2 - 1 + \frac{1}{2}, \ldots$

    _____

5. Write the following sum in summation notation ($\sum$)

    $$7 + 14 + 28 + \ldots + 896$$

    _____

6. Evaluate

a.) $\sum_{n=1}^{\infty} \frac{3}{5^n} =$

b.) $\sum_{n=1}^{\infty} \frac{3^n}{5^n} =$

1. Evaluate: $_7C_2$ and $_8P_3$:

2. Write the first three terms of the expansion and simplify: $(2x - 3)^6$

3. Find the 5th term of $(a + 2b)^{10}$.

_____

4. Determine the coefficient of $x^3 y^y$ in the expansion of $(\frac{1}{4}x - 2y^2)^7$.

_____

5. Expand and simplify: $(2 - 3b)^3$

_____

6. Give the first three terms of $(a + b)^{-3}$.

_____ _____

1. Given: $x^2 = 12y + 1$.   Vertex: _____ Focus: _____

    Directrix Line: _____

    Does it go up, down, left, or right?

2. Given: $y^2 = -8x$ Vertex: _____ Focus: _____

    Directrix Line: _____

    Does it go up, down, left, or right?

3. Graph the following: $y^2 + 2y - x + 1 = 0$. Label the vertex, focus, directrix,   an d axis of symmetry.

4. Write the standard form of the equation of the parabola with Vertex:(-1,2) and F ocus: (-1,0)

5. Write the standard form of the equation of the parabola with Vertex: (2,4) and d irectrix line: (x = 4).

1. Given: $\dfrac{(x+2)^2}{16} + \dfrac{(y-3)^2}{9} = 1$. Find the following and make a sketch:

Sketch

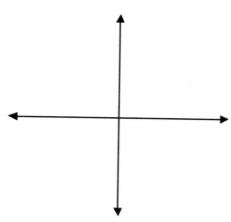

a.) Center: _____

b.) Vertices: _____

c.) Endpoints of Minor Axis: _____

d.) Length of Major Axis: _____

e.) Length of Minor Axis: _____

f.) Eccentricity= _____

g.) Foci: _____

2. Write the following in standard form and sketch. Label Center, vertices, and foci

$$x^2 + 16y^2 - 160y + 384 = 0.$$

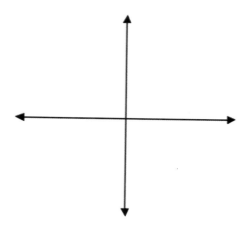

3. Write the equation of the ellipse in standard form with Foci $(0, \pm 1)$, and Vertices $(0, \pm 4)$.

PreCalculus Section 8.3 Work Sheet

Graph the below equation and indicate all important points.

1. $y = 5x^2 - 10x + 6$

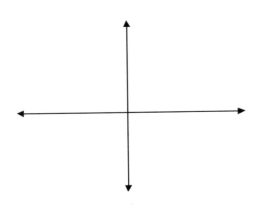

2. $x^2 - 4y^2 + 2x + 8y = 7$

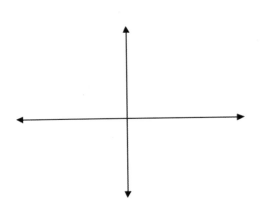

3. $16x^2 + 25y^2 - 32x - 150y = 159$

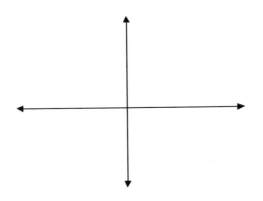

4. $x^2 + y^2 + 8x = -12$

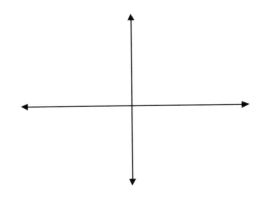

PreCalculus Section 8.4 Work Sheet

1. Sketch the graph of $\frac{(y-2)^2}{25} - \frac{(x+3)^2}{16} = 1$.   Center= _____

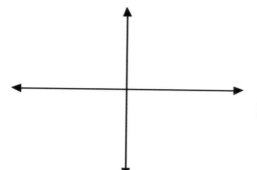

   Points: _____

   es: _____

   h of Transverse Axis: _____

Equations of the asymptotes: _____

2. Sketch the graph of $x^2 - 4y^2 + 2x + 8y = 7$. Label Center, vertices, and foci.

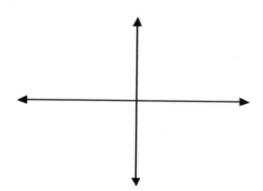

3. Write the equation of the hyperbola in standard form with Center (-1, 5);   Vertex (-1, 6);

   and Focus (-1, 9)

   _____

1. Sketch the curve represented by the parametric equation. Then, eliminate the parameter and write the corresponding rectangular equation whose graph represents the curve. Indicate the direction of the curve.

a.) $x = 1 - 3t \ and \ y = 5 + 2t$

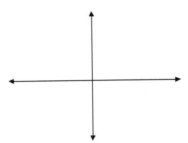

b.) $x = t \ and \ y = t^3$

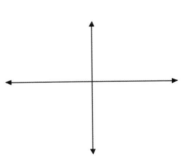

c.) $x = \sqrt{t} \ and \ y = 1 - t$

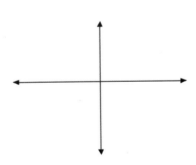

d.) $x = 4\cos t - 1 \ and \ y = 3\sin t + 2$

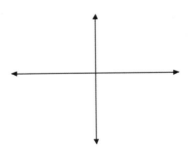

PreCalculus Section 8.6 Work Sheet

1. Plot the points and give three additional polar representations.

a.) $\left(2, -\frac{7\pi}{4}\right)$

b.) $\left(-3, \frac{5\pi}{6}\right)$

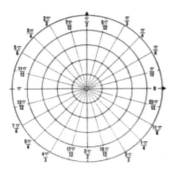

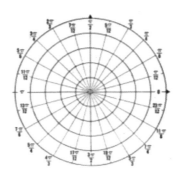

Additional Representations

Additional Representations

_____

_____

2. Convert $(5\sqrt{2}, -\frac{11\pi}{6})$ to rectangular form.

3. Convert (3, -1) to polar form.

4. Convert the rectangular equation $x^2 + y^2 - 8y = 0$ to Polar form.

5. Without sketching, describe the graphs of (a.) $r = 2$ and (b.) $r = \frac{1}{\sin\theta}$.

PreCalculus Section 8.7 Work Sheet

1. Sketch the graph of the polar equations. Fill out the table as well.

   a.) $r = 4\cos\theta$

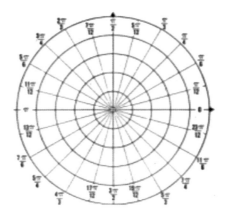

θ (degrees)	0	30°	60°	90°	120°	150°	180°	210°	270°	330°	360°
radians	0	$\dfrac{\pi}{6}$	$\dfrac{\pi}{3}$	$\dfrac{\pi}{2}$	$\dfrac{2\pi}{3}$	$\dfrac{5\pi}{6}$	π	$\dfrac{7\pi}{6}$	$\dfrac{3\pi}{2}$	$\dfrac{11\pi}{6}$	2π
r											

   b.) $r = 5 - 4\sin\theta$

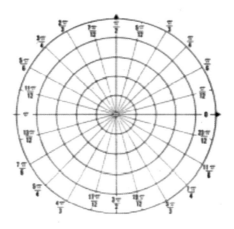

θ (degrees)	0	30°	60°	90°	120°	150°	180°	210°	270°	330°	360°
radians	0	$\dfrac{\pi}{6}$	$\dfrac{\pi}{3}$	$\dfrac{\pi}{2}$	$\dfrac{2\pi}{3}$	$\dfrac{5\pi}{6}$	π	$\dfrac{7\pi}{6}$	$\dfrac{3\pi}{2}$	$\dfrac{11\pi}{6}$	2π
r											

1. Each of the following is an equation of a conic section. State which one and find, if they exist:

(I) the coordinates of the center, (II) the coordinates of the vertices, (III) the coordinates of the foci, (IV) the eccentricity, (V) the equations of the asymptotes. Also, sketch the graph.

a.) $9x^2 - 16y^2 - 18x + 96y + 9 = 0$

b.) $4x^2 + 4y^2 - 12x - 20y - 2 = 0$

c.) $4x^2 + y^2 + 24x - 16y = 0$

d.) $y^2 + 6x - 8y + 4 = 0$

2. Find the equation of the hyperbola with center at (3, -4), eccentricity 4, and transverse axis on y axis of length 6.

3. What is the graph of each of the following?

a.) $x^2 + y^2 - 4x + 2y + 5 = 0$

b.) $xy = 0$

c.) $2x^2 - 3y^2 + 8x + 6y + 5 = 0$

d.) $3x^2 + 4y^2 - 6x - 16y + 19 = 0$

e.) $x^2 + y^2 + 5 = 0$

PreCalculus Section 9.1 Work Sheet

1. Find the limit of $\lim\limits_{x\to 4}(x^2 - 3x + 1)$ by using the table feature on the calculator.

x	3.9	3.99	3.999	4	4.001	4.01	4.1
f(x)							

Limit of f(x) as x approaches 4 is_____

2. Find the limit of $\lim\limits_{x\to 1}\dfrac{x-3}{x^2 - 4x + 3}$ by using your graphing calculator and graphing it.

Limit of f(x) as x approaches 1 is_____

3. Does the limit exist why or why not.

a.) $\lim\limits_{x\to 3}\dfrac{|x-3|}{x-3}$

b.) $\lim\limits_{x\to 0}\dfrac{2e^x - 2}{x - 2}$

c.) $\lim\limits_{x\to 0}\dfrac{\sqrt{x+5} - 4}{x - 2}$

4. Find the limits:

a.) $\lim\limits_{x\to -2}(2x^3 - 6x + 5)$ b.) $\lim\limits_{x\to -5}\dfrac{8}{x + 2}$ c.) $\lim\limits_{x\to 8}\dfrac{\sqrt{x+1}}{x - 2}$

1. Evaluate the limits:

a.) $\lim_{x \to -3} (\frac{1}{3}x^2 - 4x)$ _____

b.) $\lim_{x \to 4} \dfrac{x - 2}{x^2 + 2x + 2}$ _____

c.) $\lim_{x \to 3} \dfrac{x^2 - 5}{3x}$ _____

2. Find the limits, if they exist.

a.) $\lim_{x \to 4} \dfrac{4 - x}{x^2 - 16}$ _____

b.) $\lim_{a \to -3} \dfrac{a^3 + 27}{a + 3}$ _____

c.) $\lim_{b \to 0} \dfrac{\sqrt{7 - b} - \sqrt{7}}{b}$ _____

3. Approximate the limit to three decimal places:

a.) $\lim_{x \to 0} \dfrac{\sin x}{2x}$

          _____

1. Use the definition of Derivative to find the slope of the graph of the function at the specified point.

a.) $h(x) = 3x + 5$ at $(-1, -3)$ _____

b.) $f(x) = 2x - x^2$ at $(3, 12)$ _____

c.) $f(x) = \dfrac{2}{(x-1)}$ at $(2, 1)$ _____

2. Find the derivative of $f(x) = x^3 + 3x$. Determine where the tangent line is horizontal.

PreCalculus Section 9.4 Work Sheet

1. Find the limit (if it exists).

a.) $\lim\limits_{x \to \infty} \dfrac{3}{2 + x}$ _____

b.) $\lim\limits_{x \to \infty} \dfrac{5 - 2x}{1 + 3x}$ _____

c.) $\lim\limits_{y \to \infty} \dfrac{3y^3}{y^2 - 32}$ _____

2. Find the limits of the sequences (if it exists.) Assume n begins with 1.

a.) $a_n = \dfrac{4n - 1}{n + 3}$ _____

b.) $a_n = \dfrac{4n - 5}{3}$ _____

c.) $a_n = \dfrac{(-1)^{n+1}}{n^2}$ _____

Geometry

Geometry Review

Provides problems for review and self-diagnosis
through appropriate review questions to see if
you have a good understanding of geometry.

Triangle Review

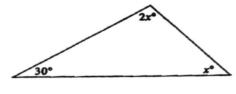

Figure 1

1. In Figure 1, what is the value of the largest angle?

(A) 30   (B) 50   (C) 60   (D) 100   (E) 120

2. What is the area of an equilateral triangle whose altitude is 9?

(A) 27   (B) $18\sqrt{3}$ (C) $27\sqrt{3}$ (D) 36   (E) $36\sqrt{3}$

3. Two sides of a right triangle are 15 and 17. Which of the following could be the length of the third side?

I. 8   II. 12   III. $\sqrt{514}$

(A) I only   (B) II only   (C) I and II only   (D) I and III only   (E) I, II, and III

4. In Figure 2, length of $AB = BC$. If the area of $\triangle ABE$ is b, what is the area of $\triangle ACD$?

(A) $b\sqrt{2}$

(B) $b\sqrt{3}$

(C) $2b$

(D) $3b$

(E) $4b$

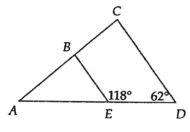

Figure 2

5. In Figure 3, $\triangle ABC$ is equilateral and $\triangle ADC$ is isosceles. If $AC = 2$, what is the distance from B to D?

(A) 0.286

(B) 0.318

(C) 0.567

(D) 0.732

(E) 1.276

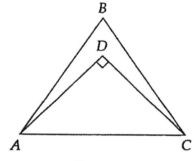

Figure 3

6. In Figure 4, what is the length of PS in the triangle?

(A) $5\sqrt{2}$

(B) 10

(C) 11

(D) 13

(E) $12\sqrt{2}$

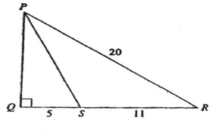

Figure 4

7. What is the value of x in the figure 5?

(A) 90

(B) 110

(C) 115

(D) 125

(E) 130

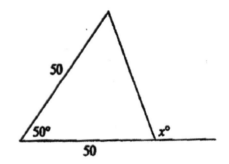

Figure 5

8. In figure 6, what is the value of w?

(A) 100

(B) 110

(C) 120

(D) 130

(E) 140

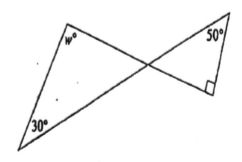

Figure 6

Questions 9 and 10 refer to the following figure

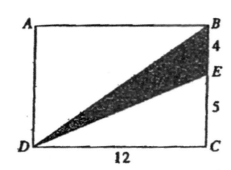

ABCD is a rectangle.

9. What is the area of $\triangle BED$?

(A) 12

(B) 24

(C) 36

(D) 48

(E) 60

10. What is the perimeter of $\triangle BED$?

(A) $19+5\sqrt{2}$

(B) 28

(C) $17+\sqrt{185}$

(D) 32

(E) 36

Quadrilaterals and Other Polygons Review

1. A regular hexagon and a square have the same perimeter. If the perimeter of the square is 12, what is the area of the hexagon?

   (A) 9.000
   (B) 10.392
   (C) 11.352
   (D) 13.856
   (E) 13.5

2. In the figure below, the perimeter of isosceles trapezoid $ABCD$ is 50. If $BC = 9$ and $AD = 21$, what is the length of diagonal AC?

   (A) 13
   (B) 14
   (C) 15
   (D) 16
   (E) 17

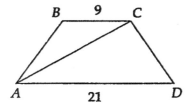

3. In the figure at the right, the two diagonals divide square ABCD into four small triangles. What is the sum of the perimeters of all <u>those triangles</u>?

   (A) $2+2\sqrt{2}$

   (B) $8+4\sqrt{2}$

   (C) $8+8\sqrt{2}$

   (D) 16

   (E) 24

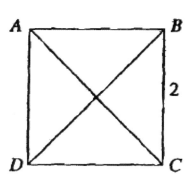

4. If the length of a rectangle is 4 times its width, and if its area is 144, what is its perimeter?

(A) 6

(B) 24

(C) 30

(D) 60

(E) 96

5. If the angles of a five-sided polygon are in the ratio of 2:3:4:4:5, what is the degree measure of the smallest angle?

(A) 30

(B) 40

(C) 60

(D) 80

(E) 90

Questions 6 and 7 refer to a rectangle in which the length of each diagonal is 18, and one of the angles formed by the diagonal and a side measures 30°.

6. What is the area of the rectangle?

(A) 27

(B) 108

(C) $36\sqrt{3}$

(D) $81\sqrt{2}$

(E) $81\sqrt{3}$

7. What is the perimeter of the rectangle?

(A) 36

(B) 81

(C) $18+18\sqrt{3}$

(D) $27+9\sqrt{3}$

(E) $81\sqrt{2}$

8. The length of a rectangle is 5 more than the side of a square, and the width of the rectangle is 5 less than the side of the square. If the area of the square is 45, what is the area of the rectangle?

(A) 20   (B) 25   (C) 45   (D) 50   (E) 70

Questions 9 and 10 refer to the following figure, in which M and N are midpoints of two of the sides of square ABCD. .

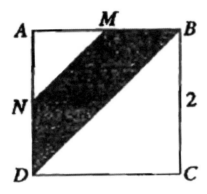

9. What is the perimeter of the shaded region?

(A) 3
(B) $2+3\sqrt{2}$
(C) $3+2\sqrt{2}$
(D) 5
(E) 8

10. What is the area of the shaded region?

(A) 1.5
(B) 1.75
(C) 3
(D) $2\sqrt{2}$
(E) $3\sqrt{2}$

Circle Review

1. In the figure below, if the radius of circle O is 8, what is the length of diagonal $\overline{AC}$ of rectangle OABC?

   (A) $\sqrt{2}$

   (B) $\sqrt{8}$

   (C) $4\sqrt{2}$

   (D) 8

   (E) $8\sqrt{2}$

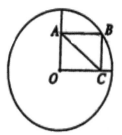

2. In the figure right, points A, B, and C lie on the circumference of the circle centered at O. If $\angle OAB$ measures 50° and $\angle BCO$ measures 60°, what is the measure of $\angle AOC$?

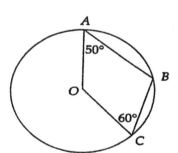

   (A) 110°

   (B) 120°

   (C) 130°

   (D) 140°

   (E) 150°

3. What is the circumference of a circle whose area is 36π?

   (A) 6

   (B) 12

   (C) 6π

   (D) 12π

   (E) 18π

4. What is the area of a circle whose circumference is π?

    (A) $\frac{\pi}{4}$

    (B) $\frac{\pi}{2}$

    (C) π

    (D) 2π

    (E) 4π

5. What is the area of a circle that is inscribed in a square of area 4?

    (A) $\frac{\pi}{4}$

    (B) $\frac{\pi}{2}$

    (C) π

    (D) $\sqrt{2}\pi$

    (E) 2π

6. A square of area 2 is inscribed in a circle. What is the circumference of the circle?

    (A) $\frac{\pi}{4}$ (B) $\frac{\pi}{2}$ (C) π (D) $\sqrt{2}\pi$ (E) 2π

7. In the figure above, what is the value of x?

    (A) 30

    (B) 36

    (C) 45

    (D) 54

    (E) 60

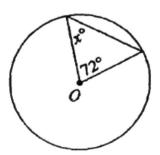

8. If A is the area and C is the circumference of a circle, which of the following is an expression for area in terms of C?

(A) $\frac{C^2}{4\pi}$ (B) $\frac{C^2}{4\pi^2}$ (C) $2C\sqrt{\pi}$ (D) $2C^2\sqrt{\pi}$ (E) $\frac{C^2\sqrt{\pi}}{4}$

9. What is the area of a circle whose radius is the diagonal of a square whose area is 1?

(A) 2

(B) $2\sqrt{2}\pi$

(C) 4π

(D) 8π

(E) 16π

10. In the figure above, l is tangent to circle O at A, and OA = AB = 2. What is the area of the shaded region?

(A) $\frac{\pi}{2}$

(B) $4 - \frac{\pi}{2}$

(C) $2 - \frac{\pi}{2}$

(D) $2 - \pi$

(E) $4 - 4\pi$

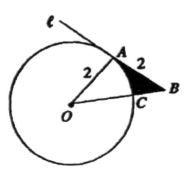

Coordinate Geometry

1. In the figure below, what is the value of *a*?

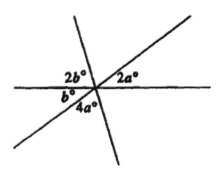

(A) 9

(B) 18

(C) 27

(D) 36

(E) 45

2. If points (4, 0), (0, 0), (0, 2), and (*a*, 2) are consecutive vertices of a trapezoid of area 9, what is the value of *a*?

(A) 1.5

(B) 2

(C) 2.5

(D) 5

(E) 9

3. In the figure below, what is the value of *x* if *y*:*x* = 4 : 1?

(A) 18

(B) 27

(C) 36

(D) 45

(E) 54

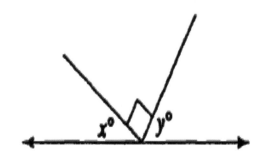

4. What is the measure of the angle formed by the minute and hour hands of a clock at 2:50?

    (A) 90°   (B) 120°   (C) 145°   (D) 150°   (E) 155°

5. Concerning the figure below, if $a = b$, which of the following statements must be true?

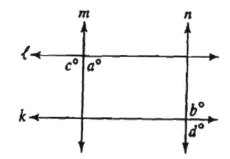

    I. $c = d$

    II. l and k are parallel.

    III. m and k are perpendicular.

(A) None

(B) I only

(C) I and II only

(D) I and III only

(E) I, II, and III

6. In the figure below, B and C lie on the line n, line m bisects $\angle AOC$, and line l bisects $\angle AOB$. What is the measure of $\angle DOE$?

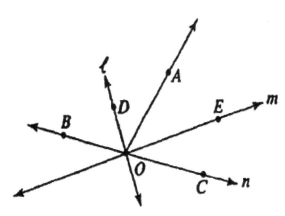

(A) 75

(B) 90

(C) 105

(D) 120

(E) It cannot be determined

    the information given.

7. If the line $y = 6x - 18$ intersects the line $y = mx + 12$ in the fourth quadrant, which of the following must be true?

   (A) $m < -4$
   (B) $-4 < m < 3$
   (C) $-3 < m < 3$
   (D) $0 < m < 4$
   (E) $m > 3$

8. The graph of the circle equation $x^2 + y^2 = 25$ includes how many points (x, y) in the coordinate plane where x and y are both integers?

   (A) 4
   (B) 5
   (C) 8
   (D) 10
   (E) 12

9. The shaded portion of Figure right shows the graph of which of the following?

   (A) $x\,(y - 2x) \leq 0$
   (B) $x\,(y - 2x) \geq 0$
   (C) $x\left(y + \frac{1}{2}x\right) \geq 0$
   (D) $x\left(y + \frac{1}{2}x\right) \leq 0$
   (E) $x\left(y - \frac{1}{2}x\right) \leq 0$

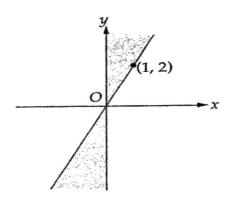

10. Which of the following equations describes the set of all points (x, y) in the coordinate plane that are a distance of 4 from the point $(-3, 4)$?

(A) $(x + 3) + (y - 4) = 4$

(B) $(x - 3) + (y + 4) = 4$

(C) $(x + 3)^2 + (y - 4)^2 = 4$

(D) $(x + 3)^2 + (y - 4)^2 = 16$

(E) $(x - 3)^2 + (y + 4)^2 = 16$

Algebra2

Algebra2 Review

We provide questions for review and self-diagnosis through appropriate review questions to see if you have a good understanding of Algebra.

1. If $3x+12 = 36$, what is the value of $x+3$?

    (A) 3
    (B) 6
    (C) 9
    (D) 11
    (E) 18

2. If $5x+13=7-1x$, what is the value of x?

    (A) $\frac{-10}{3}$
    (B) -3
    (C) -1
    (D) 1
    (E) $\frac{10}{3}$

3. If $x = y^2$ and $y = \frac{5}{c}$, what is the value of x when $c = \frac{1}{2}$?

    (A) 2.50
    (B) 3.16
    (C) 6.25
    (D) 20.00
    (E) 100.00

4. For all $w\, x\, y \neq 0$, $\frac{6x^2y^{12}w^6}{(2x^2yw)^3} =$

    (A) $\frac{y^9w^2}{x^3}$
    (B) $\frac{y^9w^4}{x^4}$
    (C) $\frac{y^9w^3}{2x^3}$
    (D) $\frac{3y^4w^2}{4x^3}$
    (E) $\frac{3y^9w^3}{4x^4}$

5. When $2x^3 + 3x^2 - 4x + c$ is divided by $x + 2$, the remainder is 3. What is the value of c ?

   (A) –1

   (B) 1

   (C) 2

   (D) 3

   (E) 5

6. If ax-b= c-dx, what is the value of x in terms of a, b , c and d?

   (A) $\frac{b+c}{a+d}$

   (B) $\frac{c-d}{a-d}$

   (C) $\frac{c-d}{a+d}$

   (D) $\frac{b+c-d}{a}$

   (E) $\frac{c}{a} - \frac{d}{b}$

7. If $\frac{1}{3}x + \frac{1}{6}x + \frac{1}{9}x = 33$, what is the value of x?

   (A) 3

   (B) 18

   (C) 27

   (D) 54

   (E) 72

8. If $16-2\sqrt{x}=13$, what is the value of x?

   (A) $\frac{9}{4}$

   (B) $\frac{29}{4}$

   (C) 36

   (D) 196

   (E) There is no value of x that satisfies the equation.

9. If $32^{a+b} = 16^{a+2b}$, then a =

(A) b

(B) 2b

(C) 3b

(D) b+2

(E) b-2

10. If the average (arithmetic mean) of 3a and 4b is less than 50, and a is twice b, what is the largest integer value of a?

(A) 9

(B) 10

(C) 11

(D) 19

(E) 20

11. If $x = 3a+8$ and $y = 9a^2$, what is y in terms of x?

(A) $(x-8)^2$

(B) $3(x-8)^2$

(C) $\frac{(x-8)^2}{3}$

(D) $\frac{(x+8)^2}{3}$

(E) $(x+8)^2$

12. Which of the following is a solution of $4|x + 1| - 5 = -1$

(A) -2

(B) 1

(C) $\frac{4}{3}$

(D) 2

(E) The equation has no solution

13. For all $x \neq \pm 3$, $\dfrac{3x^2 - 11x + 6}{9 - x^2} =$

   (A) $\dfrac{2-3x}{x-3}$

   (B) $\dfrac{2-3x}{x+3}$

   (C) $\dfrac{2x-3}{x-3}$

   (D) $\dfrac{3x-2}{x+3}$

   (E) $\dfrac{3x-2}{x-3}$

14. If $\sqrt[3]{8x + 6} = -3$, what is the value of x?

   (A) -4.125

   (B) -2.625

   (C) -1.875

   (D) -1.125

   (E) 2.625

15. If $\dfrac{5}{x+3} = \dfrac{1}{x} + \dfrac{1}{2x}$, what is the value of x?

   (A) $\dfrac{3}{14}$

   (B) $\dfrac{1}{3}$

   (C) $\dfrac{6}{13}$

   (D) $\dfrac{3}{4}$

   (E) $\dfrac{9}{7}$

1. Jenny plans to visit the Boston Museum of Art once each month in 2020 except in July and August, when she plans to go 3 times each month. A single admission costs $3.50, a pass valid for unlimited visits in any 3-month period can be purchased for $18, and an annual pass costs $60.00. What is the least amount, in dollars, that Jenny can spend for the number of visits she intends to make? d

    (A) 72

    (B) 60

    (C) 56

    (D) 49.5

    (E) 48

2. If $8^x = 16^{x-1}$, then $x =$

    (A) $\frac{1}{8}$

    (B) $\frac{1}{2}$

    (C) 2

    (D) 4

    (E) 8

3. If $a = \frac{b+x}{c+x}$, what is the value of x in terms of a, b, and c?

    (A) $\frac{a-bc}{a-1}$

    (B) $\frac{b-ac}{a-1}$

    (C) $\frac{a+bc}{a+1}$

    (D) $\frac{ac+b}{a+1}$

    (E) $\frac{ac-b}{a}$

4. If $x - 9y = 11$ and $2x + 12y = -8$, what is the value of $x + y$?

    (A) $-\dfrac{29}{11}$

    (B) $-\dfrac{9}{11}$

    (C) 1

    (D) $\dfrac{30}{6}$

    (E) $\dfrac{29}{11}$

5. Which of the following is the solution set of $|2x - 5| < 9$?

    (A) $\{x: -5 < x < 2\}$

    (B) $\{x: -5 < x < 7\}$

    (C) $\{x: -2 < x < 7\}$

    (D) $\{x: x < -5 \text{ or } x > 2\}$

    (E) $\{x: x < -2 \text{ or } x > 7\}$

6. If $\sqrt[4]{\dfrac{2x+1}{2}} = \dfrac{1}{3}$, then $x =$

    (A) -0.769

    (B) -0.488

    (C) 0

    (D) 0.488

    (E) 0.769

7. If $\dfrac{39}{5x+17} = \dfrac{39}{31}$, then $x =$

    (A) 0.4

    (B) 1.4

    (C) 2.8

    (D) 3.4

    (E) 3.8

8. If $(3^{x^2})(9^x)(9) = 81$ and $x > 0$, what is the value of x ?

(A) 0.267

(B) 0.413

(C) 0.732

(D) 1.413

(E) 1.465

9. If $y \neq 4b$, and $x = \frac{y+b^2}{y-4b}$, what is the value of y in terms of b and x?

(A) $\frac{4b-4b^2x}{x+1}$

(B) $\frac{b^2-4bx}{x+1}$

(C) $\frac{b^2+4bx}{x+1}$

(D) $\frac{b^2+4bx}{x-1}$

(E) $\frac{b^2-4bx}{x-1}$

10. If $x = 5 - 2y^2$ and $y = -2$, what is the value of x ?

(A) -2

(B) -3

(C) $- 8$

(D) 5

(E) 20

11. For all x, $2^x + 2^x + 2^x + 2^x =$

(A) 2^{x+2}

(B) 2^{x+4}

(C) 8^x

(D) 2^{4x}

(E) 8^{4x}

12. For all $x \neq \pm\frac{1}{2}$, $\frac{6x^2-x-2}{4x^2-1}=$

   (A) $\frac{2-3x}{2x+1}$

   (B) $\frac{3x+2}{2x+1}$

   (C) $\frac{3x+2}{2x-1}$

   (D) $\frac{3x-2}{2x+1}$

   (E) $\frac{3x-2}{2x-1}$

13. When $8x^3 - 5x + 9$ is divided by $x + 1$, the remainder is:

   (A) –5

   (B) –3

   (C) 1

   (D) 3

   (E) 6

14. If one of the following choices is the possible solution to the pair of equations $4x + py = 15$ and $x - py = -25$, which one is it?

   (A) $x = -3$ and $y = -5$

   (B) $x = -2$ and $y = 3$

   (C) $x = 0$ and $y = -2$

   (D) $x = 2$ and $y = 3$

   (E) $x = 3$ and $y = 5$

15. How many integers are in the solution set of $|3x + 4| < 5$?

   (A) None

   (B) Two

   (C) Three

   (D) Four

   (E) Infinitely many

Answer Sheet

PRECALCULUS

MASTER

**GEOMETRY
ALGEBRA**

REVIEW

Joe Sung

좋은땅

Complete Guide

to High School Math

MASTER PRECALCULUS

WITH GEOMETRY & ALGEBRA REVIEW

Answer Sheet

WRITTEN BY JOE SUNG

Edited by Demi_Oh

Designed by Demi_Oh

Illustrated by Demi_Oh

DEDICATED TO

My lovely wife Demi Oh

My precious Son & Daughter Ryan & Skylar Sung

My devoted parents Peter & Susana Sung

My thankful mother-in-law Sook Hyun Ki

My reliable brother Robin Sung

WRITTEN BY JOE SUNG

Edited by Demi_Oh

Designed by Demi_Oh

Illustrated by Demi_Oh

Table of Contents

PreCalculus

Table of Contents

WorkSheet

Table of Contents

Table of Contents

Geometry

Algebra 2

PreCalculus

Answer Sheet

PreCalculus Notes

This is the main section of this book, and you can learn important parts of PreCalculus by yourself. It is designed to minimize unnecessary problems and study based on the school Textbook, so you can study effectively in preparation for school.

Lecture 1.1 Functions

1. Is the relationship a function?

a.)

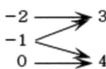

b.) {(2,1), (3,4), (4,1), (5,2), (6.5)}

all x values are different, so it is function.

-1 has 2 y values, Not a function.

c.) {(2,0), (3, 1), (2,3)}   2 has 2 different y values, so not a function.

d.)

Input Value	0	1	2	1	0
Output Value	-4	-2	0	2	4

0 and 1 has more than 1 y value, not a function.

e.) y=3x -5
Linear function is always a function, except for the vertical ones.

2. Test the following equations to see if they are functions:

a.) $2x + y^2 = 5$
    $y = \pm\sqrt{(4 - x)}$
    2 possible value of y exist for each value of x. Not a function!

b.) $x = -3y + 5$

Linear function! This is a function.

Function Notation: f(x)=y

Input	Output	Equation
x	f(x)	f(x)=8-3x

3. Evaluate the following functions: g(x)= 8 - 3x

a.) g(0)=

8-3(0) = 8

b.) $g\left(\frac{7}{3}\right) =$

8-3(7/3) = 1

c.) g(a+3)=

-1-3a

Piecewise Function: $f(x) = \begin{cases} 2x^2 + 1, x \leq 1 \\ -x^2 + 2, x > 1 \end{cases}$ condition must be considered!!

4.   a.) f(-2)=

   x=-2 so top eqn.

   2(-2)²+1= 9

b.) f(1)=

   x=1 so top eqn.

   2(1)²+1 = 3

c.) f(2)=

   x=2 so bottom eqn.

   -(2)²+2= -2

   d.) is the function continuous?

      function is not.

      continuous.

Find the Domain:

5.   a.) $f(x) = 3 - x^2$

      Domain: all real number

      Range: y≤ 3

b.) $h(x) = \dfrac{5}{x^2 - 3x}$

      so Domain

      x≠0, 3

   c.) $g(x) = \sqrt{x - 4}$

      x≥4

d.) $f(x) = \sqrt{2x} + 1$

      x≥0

Find the difference quotient and simplify:

6.   g(x)=2x+1, Find $\dfrac{g(x+h)-g(x)}{h}$, h ≠ 0

   = 2

7. $f(x) = 3x^2 - 2, find \dfrac{f(x)-f(7)}{x-7}, x \neq 7$

   $= \dfrac{(3x-147)}{(x-7)}$

Lecture 1.2 Increasing, Decreasing function, Extrema, Even and Odd

1. Which of the following graphs are graphs of functions?

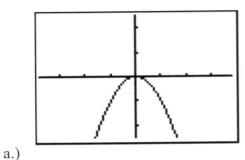

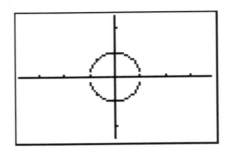

a.)   b.)

Function   Not a function.

2. Find the domain and the range of $f(x) = \sqrt{x - 3}$

Algebraic Solution   Graphical Solution (Graph it)

Domain: x-3≥0
x≥3
root function
f(x)≥0 so does
range.

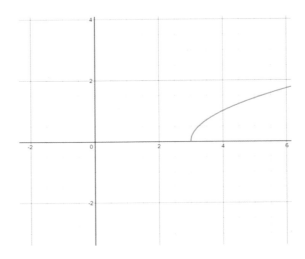

3. Determine the open intervals on which f(x) = x³-12x is increasing or decreasing.

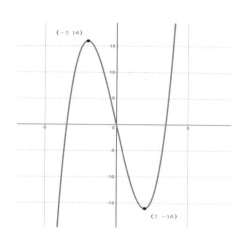

$f(x)$ is increasing on

-∞<x<-2 and 2<x<∞

$f(x)$ is decreasing on

-2<x<2

Try another question using your calculator: f(x)= 2x³+9x²

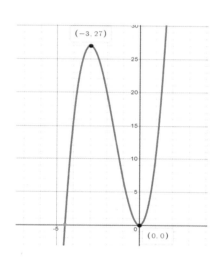

$f(x)$ is increasing on

-∞<x<-3 and 0<x<∞

$f(x)$ is decreasing on

-3<x<0

4. Use your graphing calculator to estimate the relative maximum of
$f(x) = -2x^2 - 3x + 1$.

Relative Maximum = 2.125

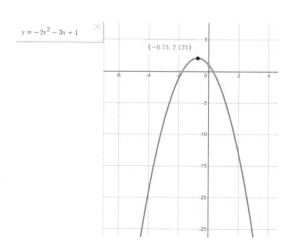

5. The profit for a shoe new company can be modeled by
$P= 0.321x^3 - 27.21x^2 + 214x + 135.7$, where P is in thousands of dollars and x is the number of units sold in thousands. What would be the maximum profit for this shoe company?

Maximum profit= $578361 would be the max. profit for the company
As you see from the graph, if you sell more than
9.44 thousand, the company will lose money.

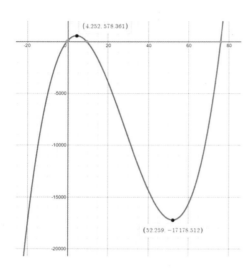

6. Graph $f(x) = |x|$

x	f(x)
-3	3
-2	2
-1	1
0	0
1	1
2	2
3	3
4	4

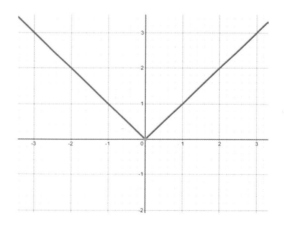

7. Sketch the graph of $f(x) = \begin{cases} x^2 + 2, x < 0 \\ 2x + 1, x \geq 0 \end{cases}$

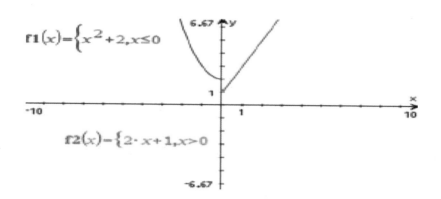

8. Test the following functions for being odd or even. Use both the algebraic and graphical approach.

Algebraic Solution   Graphical Solution

a.) $f(x) = x^4 - |x|$
   $= (-x)^4 - |-x|$
   $= x^4 - |x|$
   so even.

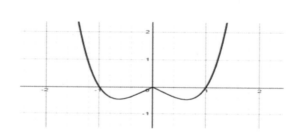

symmetric about y.

b.) $g(x) = \dfrac{x}{x^2+1}$

   $= -x/((-x)^2+1)$
   $= -[x/(x^2+1)]$
   so odd.

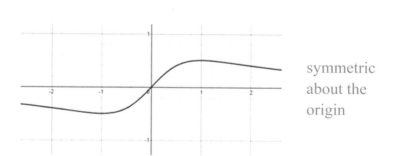

symmetric about the origin

c.) $h(x) = 2x^3 + 4x - 3$
   $= 2(-x)^3 - 4x - 3$
   $= -2x^3 - 4x - 3$
   $=$ Neither even
      nor odd.

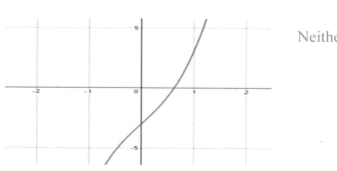

Neither!.

Lecture 1. 3 Shifting, Reflecting and Stretching Graphs

1. Graph the following common functions.

a.) $f(x) = 2$

b.) $f(x) = x$

c.) $f(x) = |x|$

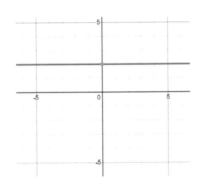

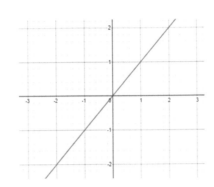

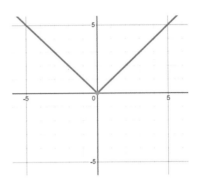

d.) $f(x) = \sqrt{x}$

e.) $f(x) = x^2$

f.) $f(x) = x^3$

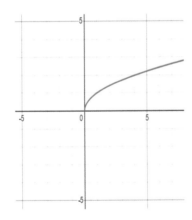

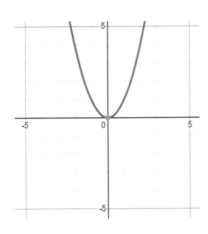

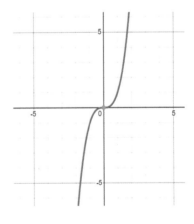

2. $f(x) = x^2$. Describe the shifts in the graph of f generated by the following functions.

a.) $g(x) = x^2 - 3$ down 3 units

b.) $h(x) = (x - 2)^2 + 5$ right 2 units and up 5 units

c.) $j(x) = (x - 6)^2$ move right 6 units

3. $f(x) = 2x^3 + 7$. Describe the reflections in the graph of f generated by the following functions:

a.) $g(x) = -2x^3 + 7$ reflect f(x) over the y=7

b.) $h(x) = -2x^3 - 7$ reflect f(x) over the y-axis.

4. Sketch the graphs of the four functions by hand on the same rectangular coordinate system.

a.) $f(x) = (x - 2)^2$

b.) $g(x) = (x - 2)^2 + 2$

c.) $h(x) = -(x - 2)^2 + 4$

d.) $j(x) = -(-x - 2)^2$

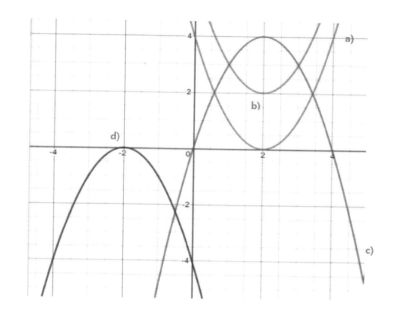

5. Graph the following

a.) $f(x) = |x - 1|$ b.) $h(x) = 3|x - 1|$ c.) $g(x) = \frac{1}{3}|x - 1|$

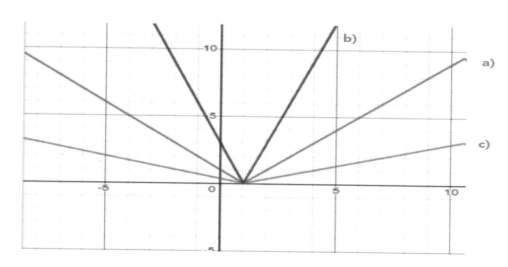

6. Write the equations of the shifts of the following common functions:

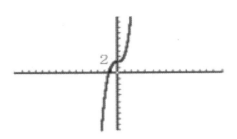

a. $y=x^3+2$

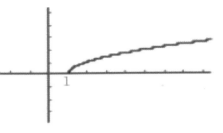

b. $y=\sqrt{(x-1)}$

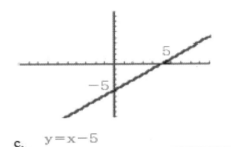

c. $y=x-5$

d. $y=|x-3|$

Lecture 1. 4 Combinations of Functions

1. $f(x) = 3x$ and $g(x) = \sqrt{x - 2}$. Find the following and specify the domain.

a.) $(f+g)(x) = 3x + \sqrt{x - 2}$ domain= $x \geq 2$

b.) $(f\text{-}g)(x) = 3x - \sqrt{x - 2}$ domain= $x \geq 2$

c.) $(fg)(x) = 3x\sqrt{x - 2}$ domain= $x \geq 2$

d.) $\left(\frac{f}{g}\right)(x) = \frac{3x}{\sqrt{x-2}}$ domain= $x > 2$

e.) $\left(\frac{2f}{g}\right)(x - 1) = \frac{6(x-1)}{\sqrt{x-3}}$ domain= $x > 3$

f.) Evaluate the following using the same f and g.

$(f+g)(3)$ $(fg)(4)$ $(f\text{-}g)(2)$

$= 10$ $= 12\sqrt{2}$ $= 6$

2. Given $f(x) = x^3 + 2$ and $g(x) = \frac{1}{x-1}$ find the following:

a.) $f \circ g$ b.) $g \circ f$

$= \frac{1}{(x-1)^3} + 2$ $= \frac{1}{(x^3+1)}$

Domain= all real number except $x = 1$ Domain= all real except $x = -1$

3. Find two functions f and g such that $(f \circ g)(x) = h(x)$.

a.) $h(x) = (2 - x)^2$

$f(x) = x^2$

$g(x) = 2-x$

b.) $h(x) = \sqrt{5 - x}$

$f(x) = \sqrt{x}$

$g(x) = 5-x$

4. Use the graphs of f and g to evaluate the functions: (assume 1 interval = 1 unit)

graph of $f(x)$

graph of g(x)

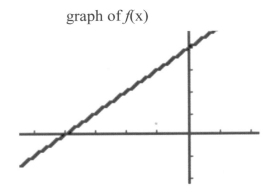
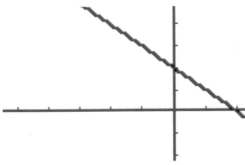

a.) $(f+g)(-2) = 6$

b.) $\left(\frac{f}{g}\right)(-1) = 1$

c.) $(fg)(-3) = 5$

d.) $(f \circ g)(0) = 6$

Application Example:

5. The number of bacteria in a certain refrigerated food is
$N(T) = 18T^2 - 85T + 400, 0 \le T \le 20$, where T is the temperature of the food in degrees Celsius. When the food is removed from refrigeration, the temperature is $T(t) = 3t + 4$, $0 \le t \le 4$, where t is the time (in hours). Find the following:

a.) The composite N(T(t)). What does this function represent?

N(T(t)) = 18(3t+4)²-85(3t+4)+400, this function represent number of bacteria at a certain time, t.

b.) The number of bacteria in the food when t=2 hours.

1350

c.) The time when the bacterial count reaches 2000.

t=2.69hr later.

Lecture 1. 5 Inverse Functions

1. Suppose $f = \{(2,1), (3,4), (5,2), (6,7)\}$ find $f^{-1}(x)$.

$f^{-1}(x) = \{(1.2), (4,3), (2,5), (7,6)\}$

2. Find the inverse of $f(x) = \frac{x-1}{4}$.

$f^{-1}(x) = 4x+1$

3. Verify your result for #2 by showing $f\left(f^{-1}(x)\right) = x$ and $f^{-1}(f(x)) = x$.

   <u>Verification</u>

   $\frac{(4x+1)-1}{4} = x$ $\qquad\qquad\qquad$ $4(\frac{(x-1)}{4}) + 1 = x$

4. Graph $f(x) = x^3$ and $g(x) = \sqrt[3]{x}$ and $y = x$ on the same coordinate plane. Are the two functions inverses of each other?

yes, as you see they are symmetric with respect to y=x.

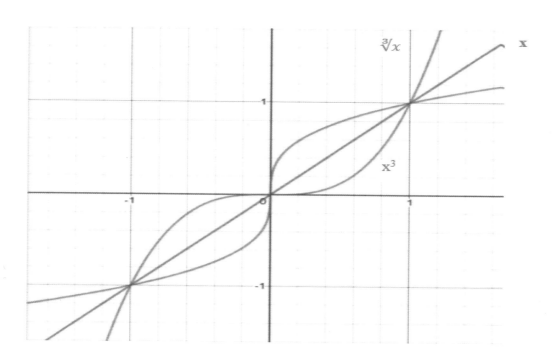

5 . Test to see if $f(x) = 2x + 3$ is one-to-one algebraically. If it is one to one, find its inverse.

Linear function is always one to one. This always produce single y value for each value of x.

x = 2y+3
x-3 = 2y
y = (x-3)/2

6. Test to see if $q(x) = 2(x - 3)^2, x \leq 3$ both algebraically and graphically. Find its inverse if you can.

x = $2(y^{-1} - 3)^2$, y≤3

$y^{-1} = -\sqrt{\frac{x}{2}}+3$, y≤3

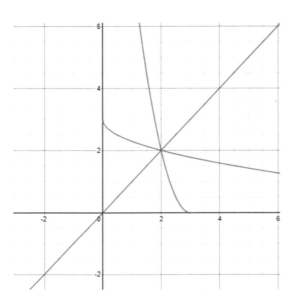

7. Find the inverse of the following functions:

a.) $f(x) = 2\sqrt[3]{x - 3}$

Inverse= $y^{-1} = \frac{x^3}{8}+3$

b.) $g(x) = \frac{x - 3}{x + 1}$

<u>Inverse</u>= $y^{-1} = \frac{x+3}{1-x}$

Lecture 2.1 Quadratic Functions

1. Graph the following function: $f(x) = (x - 1)^2 + 3$ Label the vertex, any intercepts, and the axis of symmetry.

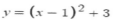
$$y = (x - 1)^2 + 3$$

Vertex= $(1, 3)$

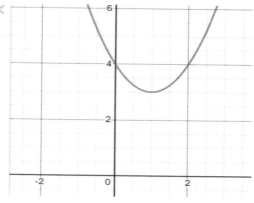

x-intercepts (set y=0)   y-intercepts(set x=0)

No x-intercept! $y = 4$

2. Sketch the graph of $g(x) = x^2 + 4x + 3$. Label the vertex, any intercepts, and the axis of symmetry.

$$y = x^2 + 4x + 3$$

vertex= $(-2, -1)$

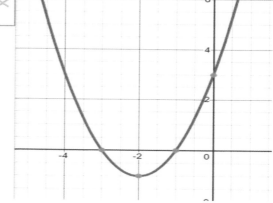

x-intercepts (set y=0)   y-intercepts(set x=0)

$(-1,0)$ $(0,3)$

$(-3,0)$

3. Write the following functions in vertex form. Identify the vertex of the following quadratic functions. Also state whether the function goes upward or downward.

a.) $h(x) = -3x^2 + 12x - 6$

$= -3(x-2)^2 + 6$

Vertex = $(2,6)$

Direction of curve: concave down

b.) $f(x) = 4x^2 + 3x - 1$

$= 4(x+\frac{3}{8})^2 - \frac{25}{16}$

Vertex = $(-3/8, -25/16)$

Direction of curve: concave up

4. Graph the following function: $f(x) = -4x^2 + 20x - 8$. Label the vertex, intercepts, and the axis of symmetry.

$x - intercept:$ 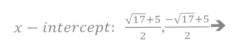 $\frac{\sqrt{17}+5}{2}, \frac{-\sqrt{17}+5}{2}$ →

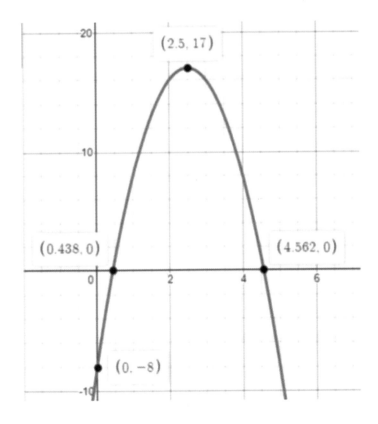

5. Find an equation for the parabola:(vertex= (-2,-9), intercepts=(-5,0) (1,0))

Equation: $y = (x + 2)^2 - 9$

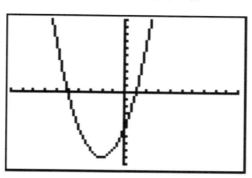

Application:

6. The path of a diver is $y = \frac{-3}{8}x^2 + \frac{24}{8}x + 10$ where y is the height (in feet) and x is the horizontal distance (in feet) from the end of the diving board. What is the maximum height of the dive? Verify your answer using a graphing utility.

max height = 16 ft.

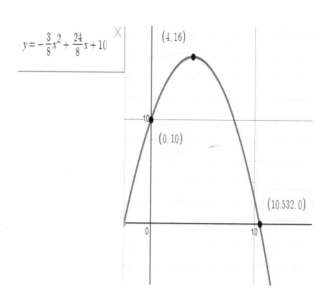

Lecture 2.2 Polynomial Functions of Higher Degree

1. Find the degree of the following polynomial functions:

a.) $f(x) = 3x^4 + 2x - 6$

Degree= 4th Deg

b.) $f(x) = 2x^3 - 3x + 5x^6$

Degree= 6th Deg

2. Label the graphs of $y = x^2, y = x^4,$ and $y = x^8$.

=> most flatten one x^8 and 2nd most x^4 and then x^2

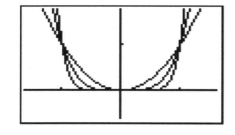

3. Label the graphs of $y = x^3, y = x^5,$ and $y = x^7$.

=> most flatten one x^7 and 2nd most x^5 and then x^3

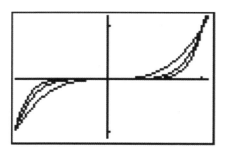

4. Sketch the graph of $f(x) = -(x + 3)^4 - 2$

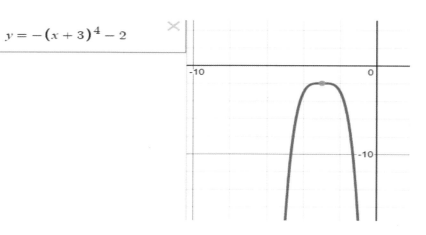

$$y = -(x+3)^4 - 2$$

5. Sketch the graph of $f(x) = (x-2)^5 + 1$

$y = (x-2)^5 + 1$

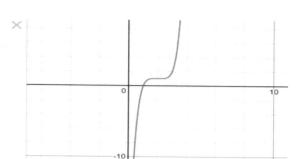

6. Describe the right-hand and left-hand behavior of the graph of each function.

a.) $f(x) = -2x^4 + 8x^3 - 14x - 2$ $\quad$ <u>both ends go $-\infty$</u>

b.) $g(x) = 3x^5 - x^3 - 12x^2 + 5$ $\quad$ <u>right end $+\infty$, left end $-\infty$</u>

7. Find the x-intercepts of the graph of $f(x) = x^3 - 2x^2 - 4x + 8$ and graph.

x intercept: -2, 2, 2

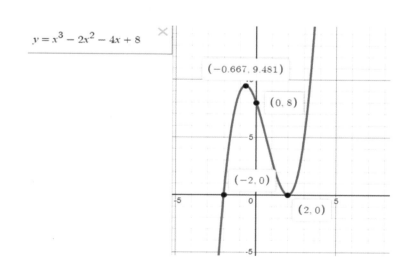

$y = x^3 - 2x^2 - 4x + 8$

$(-0.667, 9.481)$

$(0, 8)$

$(-2, 0)$

$(2, 0)$

8. Sketch the graph of $f(x) = 2x^3 - 4x^2$.

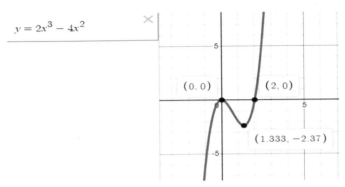

9. Find a polynomial function with the following zeros. (assume leading Coefficient =1)

a.) -2, 1, 4

$$y = (x+2)(x-1)(x-4)$$
$$= x^3 - 3x^2 - 6x + 8$$

10. Use the Intermediate Value theorem and a graphing utility to find intervals of length 1 in which the polynomial function is guaranteed to have a zero. Then approximate the zeros of the function to one-tenth.

a.) $h(x) = 28x^3 - 11x^2 + 15x - 28$

so between 0 and 1.

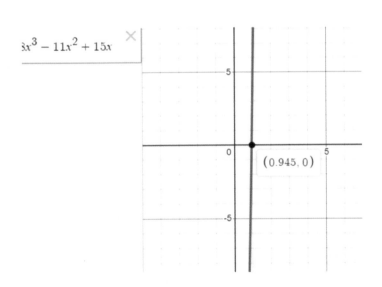

11. Match the equations with the graphs below:

a.) $y = -x^3$ b.) $y = x^2$ c.) $y = x^4$ d.) $y = -x^4$

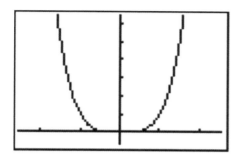

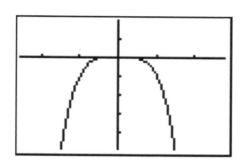

1. _____b_____   2._____a_____

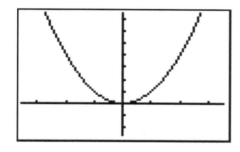

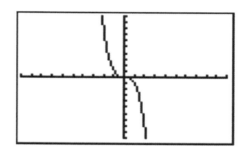

3. _____c_____   4. _____d_____

Lecture 2.3 Real Zeros of Polynomial Functions

Long Division of Polynomials

1. Divide $3x^3 - 4x^2 + x - 7$ by $x - 2$

$= 3x^2 + 2x + 5 + \dfrac{3}{(x-2)}$

2. Divide $4x^3 - x^2 + 5x - 3$ by $x^2 - 2x + 3$

$= 4x + 7 + \dfrac{(7x-24)}{(x^2-2x+3)}$

3. Use synthetic division to divide $3x^4 - 2x^2 - x + 1$ by $x - 3$

$= 3x^3 + 9x^2 + 25x + 74 + \dfrac{223}{(x-3)}$

4. Use the remainder theorem to evaluate $f(x) = 3x^2 - 8x - 2$ when $x = 3$

x = 1

5. Determine whether or not x –2 is a factor of $f(x) = x^4 - 16$ using the factor theorem.

$f(2) = 0$ so (x-2) is the factor.

6. Solve $x^3 - 2x^2 - 5x + 6 = 0$.

P: 6: $\pm 1, \pm 2, \pm 3, \pm 6$ q: 1: ± 1 $\dfrac{p}{q} = \pm 1, \pm 2, \pm 3, \pm 6$

Use synthetic division to find the solutions from the $\dfrac{p}{q}$ set.

Solutions: -2, 1, 3

7. Find all the real zeros of $2x^3 - 3x^2 - 9x + 10 = 0$.

   Zeros   -2, 1, 5/2

8. Find all the zeros: $f(x) = x^4 - 6x^3 + 6x^2 + 10x - 3$

   Zeros   1, 3, 2 – root3, 2+root3

9. Find all the zeros: $f(x) = 2x^4 - x^3 - 8x^2 - x - 10$

   Zeros   i, -i, -2, 5/2

Lecture 2.4 Complex Numbers

1. Simplify the following complex numbers by adding or subtracting:

a.) $(5 - 2i) + (4 + 3i)$

   = 9+i

b.) $(5 - i) - (2 - 4i)$

   = 3+3i

2. Multiply the following complex numbers:

a.) $3(15 - 7i) =$ 45-21i

b.) $(2 - i)(4 + 3i) =$ 8+6i-4i+3= 11+2i

c.) $(2 + 3i)^2 =$ 4+12i-9= -5+12i

d.) $(3 + 5i)(1 - 5i) =$ 3-15i+5i+25= 28-10i

3. Divide the following complex numbers:

a.) $\frac{3}{2-i} =$ (6+3i)/5

b.) $\frac{3-i}{2+3i} =$ (3-11i)/13

Plotting Complex Numbers

4. Plot each of the complex numbers in the complex plane.

a.) $4 + 2i$ b.) $5i$ c.) -3 d.) $-2 - 4i$ e.) $i - 5$

= (4,2)   = (0,5)   = (-3,0)   = (-2,-4)   = (-5,1)

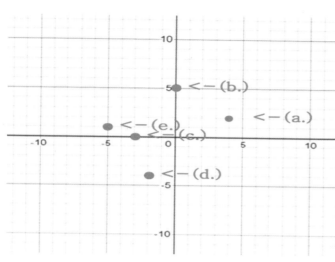

Real Axis

Imaginary Axis

5. Express each of the following powers of i as $i, -i, 1$ or -1

a.) $i^{16} =$ 1   b.) i^{25} i

c.) i^{102} -1   d.) i^{-19} i

6. Write the following in standard form:

a.) $5 + \sqrt{-28} =$ 5+2i√7   b.) $\sqrt{-100} + 4i - 5 =$ 14i-5

7. Solve for a and b.

a.) $(a + 2) + (b - 3)i = 6 + 10i$ a= 4   b= 13

Lecture 2.5 The Fundamental Theorem of Algebra

1. Solve $x^3 + 5x - 6 = 0$.

$$\left(\frac{-1+i\sqrt{23}}{2}\right), \frac{-1-i\sqrt{23}}{2})$$

Solutions_____

2. Find all real zeros of $f(x) = x^4 - 5x^3 + x - 5$

$$\left(\frac{1+i\sqrt{3}}{2}\right), \frac{1-i\sqrt{3}}{2}, -1, 5)$$

Zeros_____

3. Find a fourth-degree polynomial function with real coefficients that has 0, 2, and i as zeros.

assume leading coeff. = 1

$y = x^4 - 2x^3 + x^2 - 2x$

4. Factor $f(x) = x^4 - 12x^2 - 13$

a.) as the product of factors that are irreducible over the rationals.

$= (x^2 - 13)(x^2 + 1)$

b.) as the product of factors that are irreducible over the reals.

$= (x - \sqrt{13})(x + \sqrt{13})(x^2 + 1)$

c.) completely.

$= (x - \sqrt{13})(x + \sqrt{13})(x - i)(x + i)$

d.) List the solutions:

$= i, \ -i, \ -\sqrt{13}, \ +\sqrt{13}$

5. Find all zeros of $f(x) = x^4 + 4x^3 + 7x^2 + 16x + 12$, given that 2i is a zero.

$= 2i, -2i, -3, -1$

6. Given $f(x) = x^5 - 7x^4 + 13x^3 - 31x^2 + 36x - 12$.

Calculator question.
a.) Find all zeros of the function b.) Write the polynomial as a product of linear factors, c.) use your factorization to determine the x-intercepts of the graph of the function, and d.) use a graphing utility to verify that the real zeros are the only x-intercepts.

$= -2i, 2i, 1, 3\text{-}\sqrt{6}, 3+\sqrt{6}$

Lecture 2.6 Rational Functions and Asymptotes

1. Find the domain of $f(x) = \frac{1}{x+2}$. Then sketch a graph.

Vertical Asymptote: x = -2

Horizontal Asymptote: y = 0

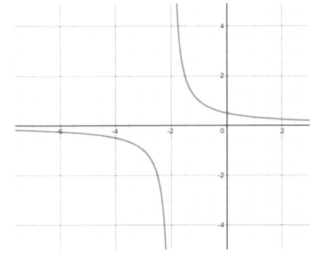

2. Given: $f(x) = \frac{2x}{x+1}$. Determine the
horizontal and vertical asymptotes and then graph.

y= 2

x= -1

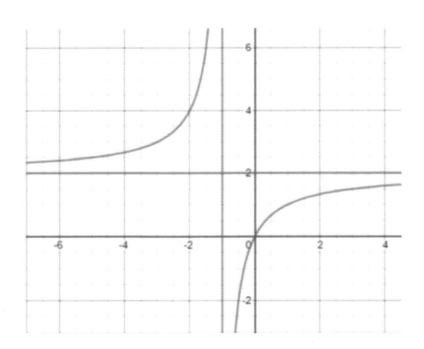

3. Given: $f(x) = \frac{3x}{x^2-9}$. Determine the horizontal and vertical asymptotes and then graph.

y=0: horizontal

x=-3: vertical

x=3: vertical

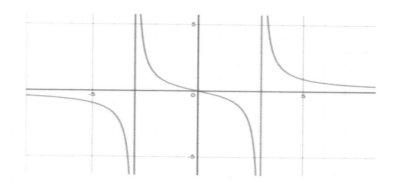

4. Given: $f(x) = \frac{x^2}{x+2}$ Determine the horizontal and vertical asymptotes and then graph.

only vertical Asymptote

at x= -2

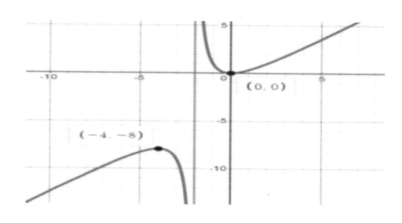

5. Given: $f(x) = \frac{x}{|x-2|}$. Determine the horizontal and vertical asymptotes and then graph.

x= 2

y= 1

y= -1

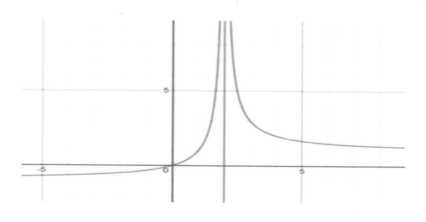

6. Given $f(x) = \dfrac{x-3}{x^2-7x+10}$ Determine the horizontal and vertical asymptotes and then graph.

x= 5

x= 2

y= 0

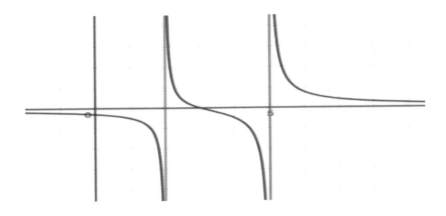

7. The Environmental Protection Organization has determined that if 800 deer are introduced to a preserve, the population at any time t (in months) is given by $N = \dfrac{800+350t}{1+0.2t}$.

What is the carrying capacity of the preserve?   1750

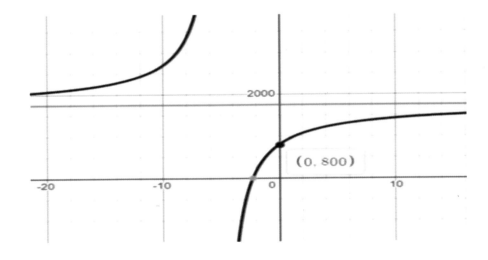

Lecture 2.7 Graphs of Rational Functions

1. Sketch the graph of $f(x) = \dfrac{1}{x-3}$. Identify the following:

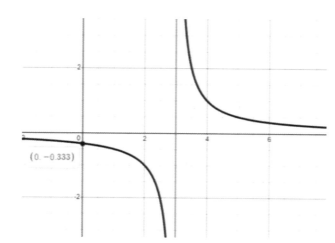

(0. −0.333)

x-intercepts: None

y-intercepts: y = -1/3

vertical asymptotes: x = 3

horizontal asymptotes: y = 0

Symmetry: only at (3,0)

Additional checking Points

x	-1	2	4	5	6
$f(x)$	$\dfrac{1}{-4}$	-1	1	$\dfrac{1}{2}$	$\dfrac{1}{3}$

2. Sketch the graph of $g(x) = \dfrac{x^2}{x^2-4}$.

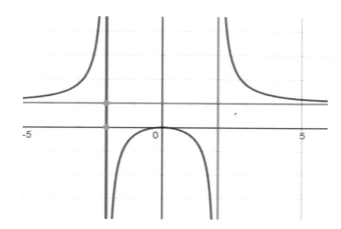

x-intercepts: x = 0

y-intercepts: y = 0

vertical asymptotes: x = -2,2

horizontal asymptotes: y = 1

Symmetry: x = 0 (y-axis)

Additional Points

x	-3	-1	-0	1	3
g(x)	9/5	-1/3	0	-1/3	9/5

3. Sketch the graph of $g(x) = \frac{x^2+2}{x-1}$.

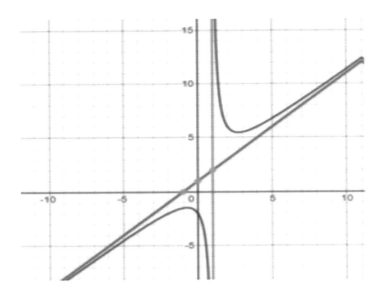

x-intercepts: None

y-intercepts: (0, -2)

vertical asymptotes: x = 1

horizontal asymptotes: none

slant asymptotes: y = x +1

Symmetry: only at (1,2)

Additional Points

x	-4	-1	0	2	3
g(x)	-18/5	-3/2	-2	6	11/2

4. Sketch the graph of $f(x) = \frac{x^2-5x+4}{x-3}$.

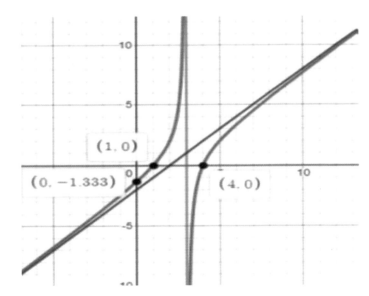

x-intercepts: (1,0), (4,0)

y-intercepts: (0,-4/3)

vertical asymptotes: x= 3

horizontal asymptotes: None

slant asymptotes: y= x-2

Symmetry: only at (3,1)

Additional Points

x	-1	0	2	5	6
f(x)	-5/2	-4/3	2	2	10/3

Lecture 3.1 Exponential Functions and Their Graphs

1. Graph the following exponential function: $f(x) = 2^x$

Domain: all real number

Range: $y \geq 0$

y-intercept: $(0,1)$

horizontal asymptote: $y = 0$

additional points:

$(-2, 1/4), (-1, 1/2), (1, 2), (2, 4), (3, 8)$

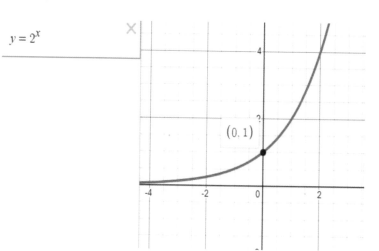

2. Graph the following exponential function: $f(x) = 2^{-x}$

Domain: all real number

Range: $y \geq 0$

y-intercept: $(0,1)$

horizontal asymptote: $y = 0$

additional points:

$(-2, 4), (-1, 2), (0,1), (1, 1/2), (2, 1/4)$

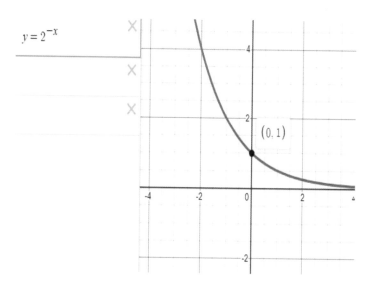

3. Graph each of the following on the same coordinate axis:

a.) $g(x) = 3^x$ b.) $f(x) = 3^{x-2}$ c.) $h(x) = 3^{x-2} + 4$

Hints:
b.) $= g(x-2)$
c.) $= g(x-2) + 3$

*Use the transformation rules

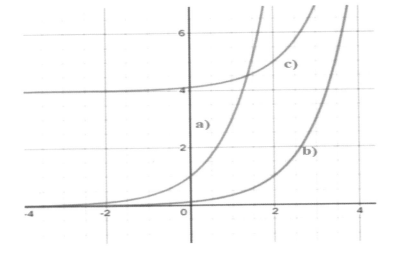

The Natural Base e
An irrational number (Euler's number) $e \approx 2.7182818\ldots$

4. Use a calculator to complete the following table:

x	1	10	100	1000	10000	100000
$\left(1 + \dfrac{1}{x}\right)^x$	2	2.5937	2.70481	2.716924	2.718146	2.71826824

Note that $\left(1 + \dfrac{1}{x}\right)^x$ as $x \to \infty$.

5. Graph $f(x) = e^x$

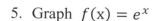

Domain: all real number

Range: $y > 0$

Intercepts: $(0,1)$

6. An investment of $4000 is made into an account that pays 2% annual interest for 10 years. Find the amount in the account if the interest is compounded:

a.) annually. n=1   b.) quarterly. n=4   c.) monthly. n=12   d.) daily, n=365

$$A(t) = A_o \left(1 + \frac{r}{n}\right)^{nt}$$

1	4	12	365
$4875.98	$4883.18	$4884.80	$4885.58

e.) What would the amount equal if interest is compounded continuously? _____

quite close but not the same.

$4885.61

7. The population of a city increases according to the model $P(t) = 29{,}000e^{.0147t}$, where t=0 corresponds to 1980. Use this model to predict the population in 2008.

Predicted population in 2008= 43,768

8. Let Q(t) (in grams) represent the mass of a quantity of carbon 14, which has a half-life of 5710 years. The quantity present after t years is $Q(t) = 50\left(\frac{1}{2}\right)^{\frac{t}{5710}}$.

a.) Determine the initial quantity (when t=0) 50g
b.) Determine the quantity present after 3000 years. 34.74g
c.) Sketch the graph of the function over the interval t=0 to t=10,000

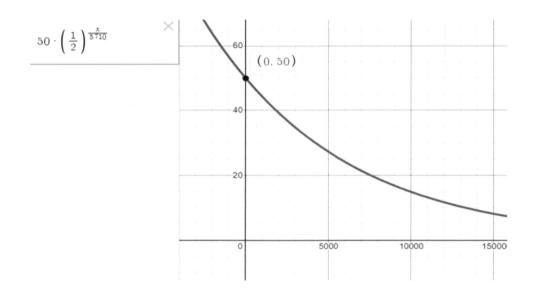

Lecture 3.2 Logarithmic Functions and Their Graphs

1. Evaluate each of the following:

a.) $log_2 4 =$

2

b.) $log_2 0.125 =$

-3

c.) $log_3 27 =$

3

d.) $log_{10} 1000 =$

3

(Note: we typically write $log_{10} 1000 =$ as $log 1000 =$ since the calculator use log base 10, it is called the common logarithm.)

e.) $log_{10}(-1.2) =$

undefined

f.) $log_7(-43) =$

undefined

2. Solve the following equations:

a.) $log_5 x = log_5 12$

12

d.) $log_4 1 = x$

0

Graphs of Exponential Functions:

3. Sketch the graph of the following on the same coordinate axis.

a.) $y = log_{10} x$

b.) $y = log_{10}(x + 3)$

c.) $y = log_{10}(x + 3) - 2$

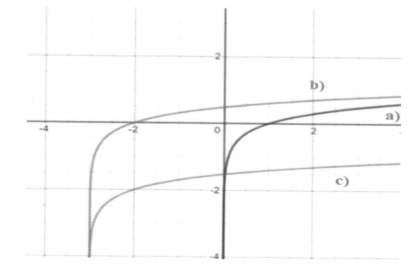

4. Evaluate the following:

a.) $\ln e^3 = 3$

b.) $e^{\ln 7} = 7$

c.) $\ln \frac{1}{e^4} = -4$

5. Use your calculator to evaluate each expression:

a.) $\ln 5.3 = 1.6677$

b.) $\ln 1.4 = 0.33647$

c.) $\ln (-3.6) = $ undefined

6. Find the domain of the following functions:

a.) $f(x) = \ln(x + 5)$

b.) $f(x) = 2 \ln|x|$

x>-5

all real number except x= 0

Application:

7. The model $t = 12.542 \ln \left(\frac{x}{x-1000} \right), x > 1000$ approximates the length of a home mortgage of $149,000 at 7.8% in terms of the monthly payment. In the model, t is the length of the mortgage in years and x is the monthly payment in dollars. Find the length of the home mortgage of $149,000 at 7.8% if the monthly payment is $2000 and the total interest charged over the life of the loan.

8.693yr

Lecture 3.3 Properties of Logarithms

1. Evaluate the following:

a.) $log_4 32 = \quad \dfrac{5}{2}$

b.) $log_3 51 = \quad 3.5789$

2. Expand the logarithmic expressions:

a.) $\log(5x^2 y^5) =$ \qquad log5+2logx+5 logy

b.) $\ln \dfrac{\sqrt{x+3}}{y^4} =$ \qquad $\frac{1}{2}$ln(x+3) - 4lny

c.) $\ln \dfrac{2x\sqrt{x-2}}{y^{4/3}} =$ \qquad ln2+ lnx+$\frac{1}{2}$ln(x-2)-$\frac{4}{3}$lny

3. Condense the logarithmic expression:

a.) $3\log x - 5\log y + \dfrac{1}{2}\log(3z) = \quad$ log ((x³+√3z)/y⁵)

b.) $\dfrac{2}{3}(3ln\ x - 5ln\ y + ln(z-5)) = \quad \frac{2}{3}$ (ln(x³ (z-5))/y⁵)

$$= \ln((x^2(z-5)^{2/3})/y^{10/3})$$

c.) $\dfrac{1}{5}(4ln\ x + 3ln\ 5 \cdot ln(z+1)) = \quad$ ln((x⁴ᐟ⁵ (z+1)^(ln5^(3/5)))

4. Rewrite the logarithm as a multiple of a) common logarithm and b.) a natural logarithm

$log_3 x$ \qquad a.) $\dfrac{logx}{log3}$ \qquad b.) $\dfrac{lnx}{ln3}$

5.) Solve the following problems:

a.) $(log_2 x)^2 = 16$

$= 16, \dfrac{1}{16}$

b.) $log_2 x^2 = 16$

$= -256, 256$

c.) $log_5(x-3) + log_5(x+3) = 4$

$= \sqrt{634}$

d.) $3^{2x} = 243$

$= \dfrac{5}{2}$

e.) $8^{(2x-1)} = 4^{(x-3)}$

$= -\dfrac{3}{4}$

f.) $3^{(2x-4)} = 12^{(3x+1)}$

$= -1.308485$

g.) $7^{(3x-2)} = \ln2 \cdot 7^{(2x+5)}$

$= 6.81164962$

Lecture 3.4 Solving Exponential and Logarithmic Equations

1. Solve each equation and round your answer to three decimal places.

a.) $4e^{2x} = 16$

$x = \ln\frac{4}{2}$

b.) $5e^{x+2} - 8 = 14$

$x = -.518396$

c.) $2(3^x - 1) = 10$

$x = 1.63093$

d.) $e^{2x} - e^x - 20 = 0$

$x = \ln 5$

2. Solve the following logarithmic equations and round your answer to three decimal places.

a.) $3\log x = 5$

$x = 10^{5/3}$

b.) $\ln\sqrt{x + 3} = \ln x$ (check for extraneous solutions)

$x = \dfrac{\sqrt{13}+1}{2}$ => one answer

c.) $\log x - \log(x - 2) = 1$

$x = \dfrac{20}{9}$

3. Approximate the solution of $2\ln(x - 1) = -x^2 + 9$. Graph both sides of the equation and look for points of intersection. (Use the Calculator's intersect option on the graphing calculator)

Make a sketch of the graph and write your answer to 5 decimal places.

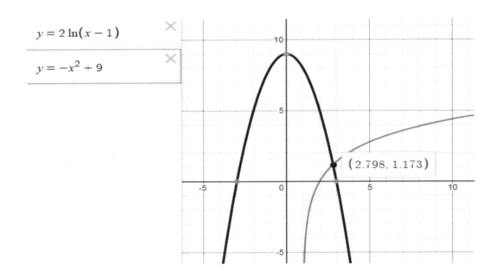

$x = 2.798$

Application

4. How long would it take for an investment to double if the interest was compounded continuously at 3.5%? $(A(t) = Ae^{rt})$

19.8042yr

5. You have $80,000 to invest. You need to have $500,000 to retire in thirty years. At what continuously compounded interest rate would you need to invest at to reach your goal?

$(A(t) = Ae^{rt})$

6.11%

6. If $500 is invested at 6%, compounded continuously, how long (to the nearest year) will it take for the money to triple?

18.31

Lecture 3.5 Exponential and Logarithmic Models

1. The population of a large city can be modeled by $y = 1.95e^{0.0226x}$, in millions, where x=0 corresponds to 1995. In what year is the population of this city expected to reach 2.8 million?

16.01 yr later

2. Two hours after bacteria were introduced to a culture the population was 100. Five hours after that the population was 400. What will the population be 24 hours after the start of the experiment? (Assume, this is an Exponential growth model)

assume this is an exp. growth.

r = 27.7259%/hr

==> 77605.

3. The radioactive isotope ^{226}Ra has a half-life of about 1619 years. If the original amount introduced was 40 grams, how much would remain after 5,000 years? (half-life is the amount of time required for one-half of an original amount to decay.)

4.703067

4. Use a graphing utility to fit a logarithmic model to the following data:

X	2	3	4	5	10	15	20
Y	3.16	4.38	5.24	5.91	8.00	9.22	10.09

Use the regression capabilities of the graphing calculator to find the logarithmic model.

y= 1.07105+3.00946 ln x

5. The table below shows the profit for a company (in millions of dollars) for the years 1990 (x=0) to 1998. Use a graphing calculator to fit an exponential model to the data, then use the model to predict the company's profit in 2005.

T	0	1	2	3	4	5	6	7	8
P	5.80	6.35	6.94	7.60	8.31	9.10	9.95	10.89	11.92

Use the regression capabilities of the graphing calculator to find the exponential model.
Exponential Model y=5.8 · 1.09417^x

Profit in 2005 = 22.37 millions of dollars

Lecture 4.1 Measure of an Angle: Angle, DMS, Radian vs Degree

1. Find a coterminal angles for each of the following: many possible answers

a.) $\frac{11\pi}{6}$ b.) $\frac{-\pi}{4}$ c.) 180°   d.)-150°   e.) 540°

$\frac{-\pi}{6}, \frac{23\pi}{6}$ $\frac{7\pi}{4}, \frac{-9\pi}{4}$ -180°, 540°   210°, -510°   180°, -180°

2. Determine the quadrant where the terminal side of each angle lies:

a.) $\frac{916\pi}{6}$ b.) $\frac{-10\pi}{3}$ c.) $\frac{97\pi}{4}$ d.) $\frac{-17\pi}{3}$

2nd 2nd 1st 1st

3. Find the complement and supplement of each of these:

a.) $\frac{4\pi}{7}$ b.) $\frac{1\pi}{8}$ c.) $\frac{1\pi}{10}$

$-\pi/17$ and $3\pi/7$ $3\pi/8$ and $7\pi/8$ $4\pi/10$ and $9\pi/10$

4. Convert to decimal degrees:

a.) 125° 12'   b.) -100° 34'

125.2°   100.566666°

5. Convert to degrees and minutes:

a.) 52.6°   b.) 284.18 °

52°36'   284°10' 48"

6. Convert from degrees to radians:

a.) 225°   b.) -120°

$(5/4)\pi$ $-2\pi/3$

7. Convert from radians to degrees:

a.) $\frac{15\pi}{6}$ b.) $\frac{-5\pi}{3}$

450° $-300°$

8. Find the area of the sector and the length of the arc subtended by a central angle of $\frac{2\pi}{3}$ radians in a circle whose radius is 6 inches.

Arc length=4π
Area = 12π

9. Find the length of the arc on a circle of radius 30 centimeters intercepted by a central angle of 60°.

Arc Length = 10π

10. Find the radian measure of the central angle of a circle of radius 22 feet that intercepts and arc of length 10 feet.

5/11 rad.

11. If a sector of a circle has an arc length of 3π inches and an area of 9π square inches, what is the length of the radius of the circle?

$r = 4$

Lecture 4.2 Unit Circle and Special Angle

1. Evaluate the six trigonometric functions at each real number:

	sin t	cos t	tan t	cot t	sec t	csc t
	y	x	$\dfrac{y}{x}$	$\dfrac{x}{y}$	$\dfrac{1}{x}$	$\dfrac{1}{y}$
$\dfrac{7\pi}{4}$	$-\sqrt{2}/2$	$\sqrt{2}/2$	-1	-1	$\sqrt{2}$	$-\sqrt{2}$
$\dfrac{5\pi}{6}$	$1/2$	$-\sqrt{3}/2$	$-\sqrt{3}/3$	$-\sqrt{3}$	$-2/\sqrt{3}$	2
$\dfrac{-2\pi}{3}$	$-\sqrt{3}/2$	$-1/2$	$\sqrt{3}$	$\sqrt{3}/3$	-2	$-2/\sqrt{3}$
$\dfrac{13\pi}{6}$	$1/2$	$\sqrt{3}/2$	$\sqrt{3}/3$	$\sqrt{3}$	$2/\sqrt{3}$	2
$\dfrac{-7\pi}{2}$	1	0	und	0	und	1

2. Use the Value of the trig function to evaluate the indicated functions.

a.) $\sin(-\theta) = 3/5$ Find   (i) $\sin \theta$ (ii) $\csc \theta$

                    $-3/5$ $-5/3$

b.) $\cos(-\theta) = \dfrac{2}{7}$ Find   (i) $\sin \theta$ (ii) $\sec \theta$

                    $\pm \sqrt{45/7}$ $7/2$

Domain of Sine and Cosine: Range of Sine and Cosine:

all real number $-1 \leq y \leq 1$

Lecture 4.3 Trigonometry Definition

Evaluate each of the six trigonometric ratios of θ (assume θ is in first Quadrant)

1. Given: Sin θ = 7/10

cosθ= $\sqrt{51}$/10

tanθ= 7/$\sqrt{51}$

cscθ= 10/7

secθ= 10/51

cotθ= $\sqrt{51}$/7

2. Given: Tan θ = 2

cosθ= 1$\sqrt{5}$

sinθ= 2/$\sqrt{5}$

cscθ= $\sqrt{5}$/2

secθ= $\sqrt{5}$

cotθ= 1/2

3. Given: csc θ = 13/12

sinθ= 12/13

cosθ= 5/13

tanθ= 12/5

cscθ= 13/12

secθ= 13/5

cotθ= 5/12

4. The given point (-3, -4) is on the terminal side of an angle in standard position. Determine the exact values of the six trigonometric functions of the angle.

q3
sinθ= -4/5
cosθ= -3/5
tanθ= 4/3
cscθ= -5/4
secθ= -5/3
cotθ= 3/4

5. The given point (5, -13) is on the terminal side of an angle in standard position. Determine the exact values of the six trigonometric functions of the angle.

q4
sinθ= -13/$\sqrt{194}$

cosθ= 5/$\sqrt{194}$

tanθ= -13/5

cscθ= -$\sqrt{194}$/13

secθ= $\sqrt{194}$/5
cotθ= -5/13

6. *Fill in the chart and Memorize it !!

θ	0°	30°	45°	60°	90°
sin	0	1/2	$\sqrt{2}/2$	$\sqrt{3}/2$	1
cos	1	$\sqrt{3}/2$	$\sqrt{2}/2$	1/2	0
tan	0	$\sqrt{3}/3$	1	$\sqrt{3}$	und
cot	und	$\sqrt{3}$	1	$\sqrt{3}/3$	0
sec	1	$2/\sqrt{3}$	$\sqrt{2}$	2	und
csc	und	2	$\sqrt{2}$	$2/\sqrt{3}$	1

7. Find the value of each trig function using the trig identities when $\tan \theta = \sqrt{3}$ (θ is in 1st Quadrant)

a) $\sec \theta = 2$

b) $\cos \theta = 1/2$

c) $\sin \theta = \frac{\sqrt{3}}{2}$

d) $\cot \theta = \frac{\sqrt{3}}{3}$

e) $\csc \theta = \frac{2\sqrt{3}}{3}$

8. Use the trig identities to transform one side of the equation into the other:

a) $\csc \theta \tan \theta = \sec \theta$

$1/\sin\theta \cdot (\sin\theta/\cos\theta)$
$= 1/\cos\theta = \sec\theta$

b) $\cot \alpha \sin \alpha = \cos \alpha$

$\cos a/\sin a \cdot \sin a$
$= \cos a$

c) $\dfrac{\csc A}{\cot A + \tan A} = \cos A$

$$\frac{\frac{1}{\sin A}}{\frac{\cos A}{\sin A} + \frac{\sin A}{\cos A}} = \frac{\frac{1}{\sin A}}{\frac{(\cos A)^2 + (\sin A)^2}{\sin A \cdot \cos A}} = \frac{\sin A \cos A}{\sin A} = \cos A$$

9. According to the safety sticker on a 30-foot ladder, the distance from the bottom of the ladder to the base of the wall on which it leans should be 35% of the length of the ladder.

a) If the ladder is in this position, what is the acute angle between the bottom of the ladder and the ground?

69.5 ft.

b) How high up the wall will the ladder reach?

28.1 ft

Lecture 4.4 Trigonometric Functions of Any Angle

1. Let (-5, 12) be a point on the terminal side of θ. Find all six trig ratios.

q2

$\sin\theta = \dfrac{12}{13}$ $\qquad$ $\csc\theta = \dfrac{13}{12}$

$\cos\theta = \dfrac{-5}{13}$ $\qquad$ $\sec\theta = -\dfrac{13}{5}$

$\tan\theta = -\dfrac{12}{5}$ $\qquad$ $\cot\theta = -\dfrac{5}{12}$

Trigonometry ASTC

When you work with trigonometry, you'll be dealing with four quadrants of a graph. The x and y axis divides up a coordinate plane into four separate sections.

ASTC is a memory-aid for memorizing whether a trigonometric ratio is positive or negative in each quadrant: [Add-Sugar-To-Coffee]

If you don't like Add Sugar To Coffee, there's other acronyms you can use such as:

All students Take Calculus.

	Quadrant 1	Quadrant 2	Quadrant 3	Quadrant 4
sine& Cosecant	+	+	-	-
Cosine& Secant	+	-	-	+
Tangent & Cotangent	+	-	+	-

2. Given: Tan θ = -8/15 and cos θ > 0. Find the remaining trig ratios.

q4,

sinθ = -8/17 $\qquad$ cscθ = 17/-8

cosθ= 15/17 $\qquad$ secθ = 17/15

cotθ= 15/-8

3. Fill in the following chart:

	Sin θ	Cos θ	Tan θ	Cot θ	Sec θ	Csc θ
π/2	1	0	und	0	und	1
π	0	-1	0	und	-1	und
3π/2	-1	-	und	0	und	-1
2π or 0	0	1	0	und	1	und

4. Find the reference angle for each of these:

a) 110°   b) 250°   c) - 315°   d) 520°

   70°   70°   45°   20°

e) 4π/3   f) 3 π/5   g)   13π/6

   2π/3   2π/5   π/6

5. In which quadrants are the trig ratios positive?

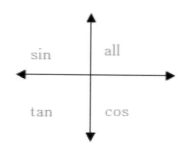

6. Evaluate the trigonometric functions for each of the following: (exact answers only)

a) cos (5 π/3)   b) tan (-240°)   c) csc 9π/4

   1/2   -√3   √2

7. Let θ be an angle in Quadrant II such that Sin θ = 1/4. Using trig identities, find the following:

a) csc θ

    4

b) tan θ

    $-1/\sqrt{15}$

c) cos θ

    $-\sqrt{15}/4$

d) sin θ · cot θ

    $-\sqrt{15}/4$

8. Use your calculator to evaluate each of the following: Round values to four decimal places:

a) tan (-200°)

    -.36397

b) cos 112°

    -.3746

c) csc (-247°)

    1.08636

d) Tan (4π/3)

    $\sqrt{3}$

e) Sin (12π/7)

    -.78183

f) Sec (15π/4)

    $\sqrt{2}$

9. Use your calculator to find <u>two</u> approximate values of θ (0° ≤ θ ≤ 360°). Round your answers to two decimal places.

a) cos θ = 0.7820

    38.56°, 321.44°

b) sin θ = 0.3880

    22.83°, 157.17°

c) tan θ = 1.7693

    60.53°, 240.525°

10. Find the indicated trigonometric value in the specified quadrant:

a) csc θ = -2, Quadrant IV, find sec θ.

$2/\sqrt{3}$

b) sec θ = -7/4, Quadrant III, find cot θ.

$4/\sqrt{33}$

c) sin θ= $\frac{2}{5}$, Quadrant II, find tan θ.

$2/-\sqrt{21}$

Lecture 4. 5 Graph of Sine and Cosine

1. Fill in the chart below

X	0	$\dfrac{\pi}{4}$	$\dfrac{\pi}{2}$	$\dfrac{3\pi}{4}$	π	$\dfrac{5\pi}{4}$	$\dfrac{3\pi}{2}$	$\dfrac{7\pi}{4}$	2π
$y=\sin x$	0	$\sqrt{2}/2$	1	$\sqrt{2}/2$	0	$-\sqrt{2}/2$	-1	$-\sqrt{2}/2$	0
$y=\cos x$	1	$\sqrt{2}/2$	0	$-\sqrt{2}/2$	-1	$-\sqrt{2}/2$	0	$\sqrt{2}/2$	1

2. Using the table above to sketch a graph of the sine and cosine functions.

a.) $y = \sin x$

Amplitude= 1

Period= 2π

Starts at $(0,0)$

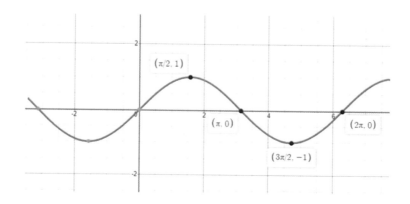

b.) $y = \cos x$

Amplitude= 1

Period= 2π

Starts at $(0,1)$

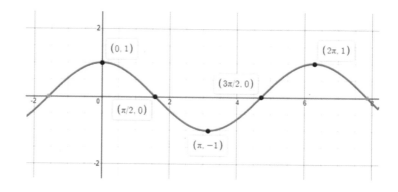

3. Graph, on the same coordinate axis, one period of each of the following functions.

a) $y = sin(x)$

b) $y = 2\ sin\ (x)$

c) $y = \frac{1}{2}sin(x)$

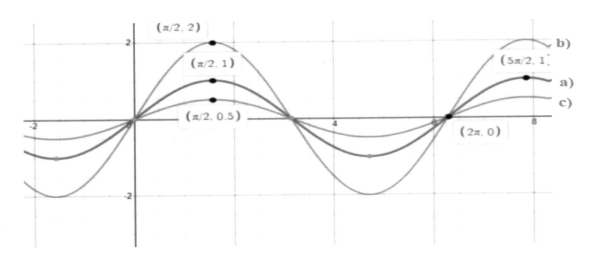

4. Graph one period of each of the following functions.

a) $y = sin3x$

b) $y = cos\left(\frac{1}{3}x\right)$

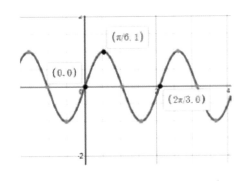

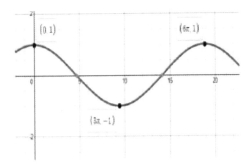

c) $y = sin\left(x - \frac{\pi}{2}\right)$

d) $y = -cos\left(\frac{\pi}{3}x\right)$

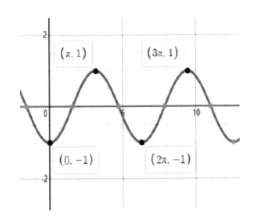

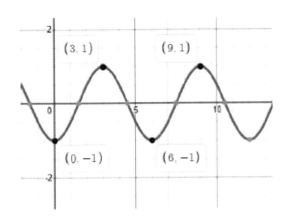

5. Find the period and amplitude for each of the following functions.

a) $f(x) = 3\sin\left(\frac{\pi}{2}x\right)$

A= 3   P= 4

b.) $g(x) = 3 + 5\sin\left(\frac{\pi}{2}x - \frac{\pi}{2}\right)$

A= 5   P= 4

c.) $h(x) = \frac{7}{3}\cos\left(\frac{3}{2}x\right)$

A= 7/3   P= 4π/3

6. Describe the relationship between the graphs f and g.

a) $f(x) = \cos x$

$g(x) = -\cos\left(x + \frac{\pi}{4}\right)$

b.) $f(x) = 2\cos x$

$g(x) = -3 + 2\cos\left(x - \frac{\pi}{2}\right)$

c.) $f(x) = \sin 3x$

$g(x) = 2 + \sin 3x$

left π/4 unit and
reflect f(x) over x-
axis.

right π/2 unit and down
3 unit.

up 2 unit.

7. Graph one period of each of the following functions

a) $f(x) = 2 + 2\sin\left(\frac{x - \pi}{3}\right)$

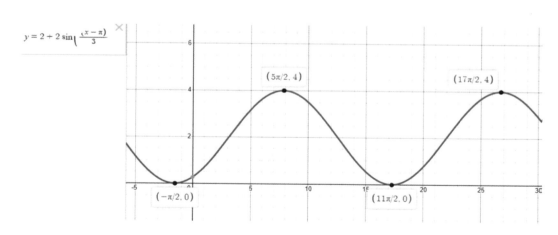

b) $g(x) = -4 + cos\left(2x - \frac{\pi}{4}\right)$

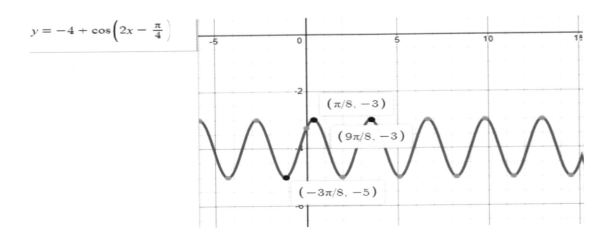

c) $f(x) = \pi \sin\left(\frac{x+1}{2\pi}\right) + 3$

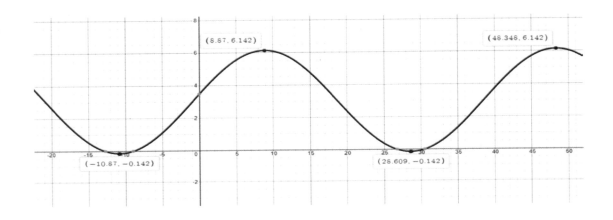

Lecture 4.6 Graph of Tangent and Cotangent

1. Fill in the chart below.

x	o	$\frac{\pi}{4}$	$\frac{\pi}{2}$	$\frac{3\pi}{4}$	π	$\frac{5\pi}{4}$	$\frac{3\pi}{2}$	$\frac{7\pi}{4}$	2π
$y=\tan x$	0	1	und	-1	0	1	und	-1	0
$y=\cot x$	und	1	0	-1	und	1	0	-1	und

2. Graph the following:

a.) $y = \tan x$

Period= π

Starts at: $(0,0)$

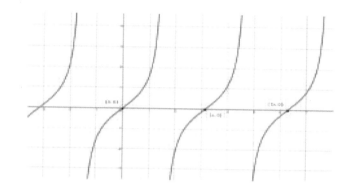

b.) $y = \cot x$

Period= π

Starts at: $(\pi/2,0)$

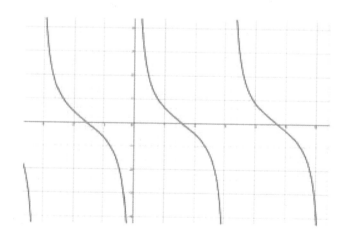

3. Give the period and graph one cycle of each of the following functions.

a) $y=\tan(\frac{1}{2}x)$

b) $y=2\cot\pi x$

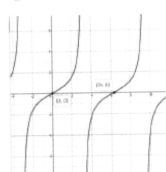

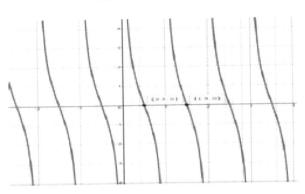

c.) $y=\cot(\pi x + \pi)$

d.) $y=2\cot\left(x + \frac{\pi}{2}\right)+1$

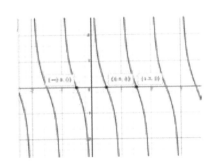

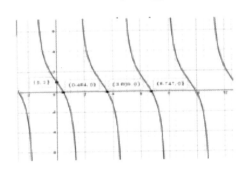

4. Graphing the sec x and csc x (First graph the sin x and the cos x)

$y=\csc x$

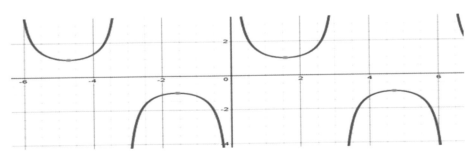

$y=\sec x$

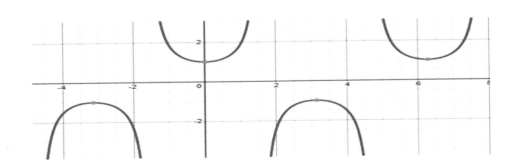

5. Graph one cycle of each of the following:

a.) $y=3\csc(2x)+1$

b.) $y=-2\sec[\frac{\pi}{4}(x-1)]$

c.) $y=2+2\csc\left(x+\frac{\pi}{4}\right)$

d.) $y=\frac{1}{2}\sec\left(\frac{\pi}{2}x+\frac{\pi}{2}\right)-3$

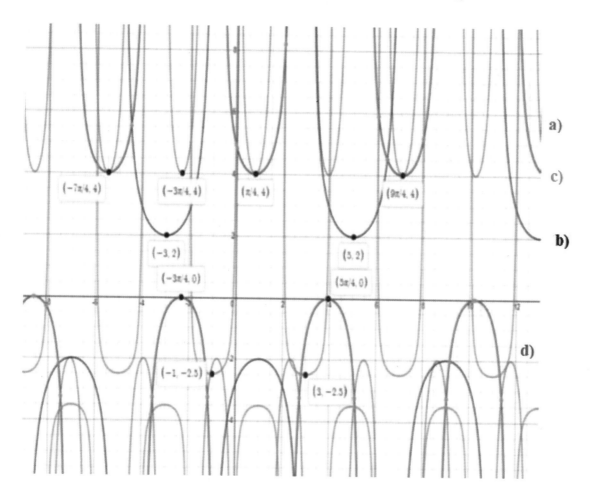

Lecture 4.7 Inverse Trigonometric Functions

1. Examine the graphs of y=sin x and y=arcsin x. (Use your calculators to draw the inverses.) Then draw the inverse like we did above.

y=sinx   y=arcsinx

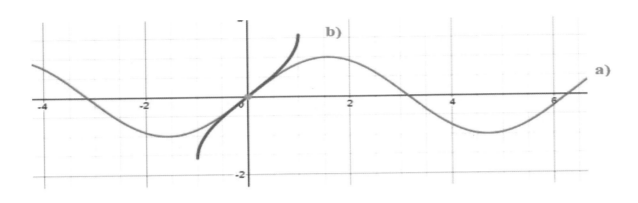

Function	Range
y=arcsin x	$\dfrac{-\pi}{2} \leq y \leq \dfrac{\pi}{2}$
y=arccos x	$0 \leq y \leq \pi$
y=arctan x	$\dfrac{-\pi}{2} < y < \dfrac{\pi}{2}$

Sketch a graph of

$y = \arctan x$ y=arccos x

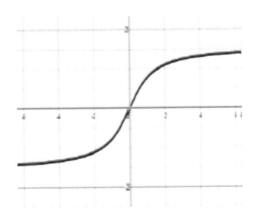
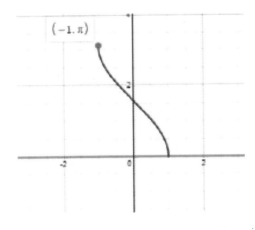

2. Without Using a calculator, find the exact values of each expression ($0° \leq \theta \leq 360°$):

a.) $\arcsin\left(\frac{-\sqrt{3}}{2}\right)$

240°, 300°

c.) $\arctan\left(\sqrt{3}\right)$

60°, 240°

b.) $\arccos\left(\frac{-\sqrt{2}}{2}\right)$

135°, 225°

d.) $\arccos\left(\frac{1}{2}\right)$

60°, 300°

3. Using a calculator, find the exact values of each expression ($0° \leq \theta \leq 360°$).:

a.) $\arcsin(-0.125)$

187.18°, 352.82°

c.) $\text{arcsec}\ (2.1)$

61.563°, 298.437°

b.) $\arctan(17.8)$

86.7845°, 266.7845°

d.) $\text{arccsc}\ (-3.5)$

196.60°, 343.40°

4. Evaluate ($0° \leq \theta \leq 360°$):

a.) $\sin(\arcsin 0.7) = 0.7$

b.) $\cos(\arccos(-0.3)) = -0.3$

5. Without using a calculator, find the exact value of the expression ($0° \leq \theta \leq 360°$).

a.) $\arcsin\left(\sin\left(\frac{\pi}{6}\right)\right) = 30°\ 150°$

b.) $\cos\left(\arcsin\left(\frac{-1}{3}\right)\right) = \pm 0.94281$

c.) $\cot\left(\arcsin\left(\frac{\sqrt{3}}{2}\right)\right) = \pm\sqrt{3}$

d.) $\cos\left(\arcsin\left(-\frac{4}{5}\right)\right) = \pm 3/5$

6. Write an algebraic expression that is equivalent to the expression ($0° \leq \theta \leq 360°$):

a.) $\sin(\arccos x) =$
$\pm\sqrt{(1-x^2)}$

b.) $\cos(\arcsin(3x)) =$
$\pm\sqrt{(1-9x^2)}$

c.) $\sec\left(\arccos\left(\frac{1}{x}\right)\right) =$

$= x$

Lecture 5.1 Fundamental Trig. Identities

1. Solve $\sin x = \sqrt{3} \cos x$ for all values of x such that $0 \leq x < 2\pi$.

$\dfrac{\pi}{3}, \dfrac{4\pi}{3}$

2. Use the values of $\sec\theta = \dfrac{-5}{2}$ and $\tan\theta > 0$ to find the values of all six trigonometric functions.

q3

$\sin\theta = \dfrac{-\sqrt{21}}{5}$ $\qquad$ $\cos\theta = \dfrac{-2}{5}$

$\tan\theta = \dfrac{\sqrt{21}}{2}$ $\qquad$ $\csc\theta = \dfrac{-5}{\sqrt{21}}$

$\cot\theta = \dfrac{2}{\sqrt{21}}$

3. Simplify: $\sin x \cdot \cos^2 x - \sin x$

$\sin x(\cos^2 x - 1) = -(\sin x)^3$

4. Simplify: $\sin x + \cot x \cdot \cos x$

$= \sin x + \dfrac{\cos x}{\sin x} \cdot \cos x$

$= \dfrac{\sin^2 x + \cos^2 x}{\sin x} = \dfrac{1}{\sin x}$

$= \csc x$

5. Factor: $\csc^2 x - \cot x - 3$

$= 1 + \cot^2 x - \cot x - 3$

$= \cot^2 x - \cot x - 2$

$= (\cot x - 2)(\cot + 1)$

6. Add and Simplify: $\dfrac{\sin\theta}{1+\cos\theta} + \dfrac{\cos\theta}{\sin\theta}$

$= \csc\theta$

7. Rewrite $\dfrac{1}{1+\sin x}$ so that it is not in fractional form.

$= \sec^2\theta - \tan\theta \cdot \sec\theta$

8. Simplify: $\dfrac{\sin\theta + \cos\theta}{\sin\theta} + \dfrac{-\cos\theta + \sin\theta}{\cos\theta}$

$= \cot\theta + \tan\theta$

9. Expand and Simplify: $(\cot x + \csc x)(\cot x - \csc x)$

$= -1$

10. Simplify: $\sin\left(\dfrac{\pi}{2} + x\right) =$

$= -\cos x$

11. Simplify: $\csc\left(\dfrac{3\pi}{2} - x\right) =$

$= -\sec x$

12. Simplify: $\tan(\pi - x) =$

$= -\tan x$

13. Simplify: $\cos(2\pi - x) =$

$= \cos x$

Lecture 5.2 Proving and Verifying trigonometric identities

Verify the following identities:

1. $\dfrac{\sin^2 x - 1}{\sin^2 x} = -\cot^2 x$

$= \dfrac{-\cos^2 x}{\sin^2 x}$

$= -\cot^2 x$

2. $\dfrac{1}{\sec x - 1} - \dfrac{1}{\sec x + 1} = 2\cot^2 x$

$= \dfrac{2}{\sec^2 x - 1}$

$= \dfrac{2}{\tan^2 x} = 2\cot^2 x$

3. $(1 + \cot^2 x)(1 - \sin^2 x) = \cot^2 x$

$= \csc^2 x \ \cos^2 x$

$= \dfrac{1}{\sin^2 x} \ \cos^2 x$

$= \cot^2 x$

4. $\sec u + \tan u = \dfrac{1}{\sec u - \tan u}$

$= \dfrac{1}{\cos x} + \dfrac{\sin x}{\cos x}$

$= \dfrac{1 + \sin x}{\cos x} \cdot \dfrac{1 - \sin x}{1 - \sin x}$

$= \dfrac{1 - \sin^2 x}{\cos x(1 - \sin x)} = \dfrac{\cos^2 x}{\cos x(1 - \sin x)}$

$= \dfrac{\cos^2 x}{\cos x(1 - \sin x)} = \dfrac{\cos x}{(1 - \sin x)}$

$= \dfrac{1}{\frac{(1 - \sin x)}{\cos x}} = \dfrac{1}{(\sec x - \tan x)}$

5. $\dfrac{\cot^2\theta}{1+\csc\theta} = \dfrac{1-\sin\theta}{\sin\theta}$

$\quad = \dfrac{\csc^2\theta - 1}{(1+\csc\theta)}$

$\quad = \csc\theta - 1$

$\quad = \dfrac{1-\sin\theta}{(\sin\theta)}$

6. $\cot t \cos t = \dfrac{1}{\tan t \sec t}$

$\quad = \dfrac{\cos t \cos t}{(\sin t)} = \dfrac{1}{\frac{\sin t}{\cos t \cos t}} = \dfrac{1}{\tan t \sec t}$

7. $(\tan^2 x + 1)(\cos^2 x - 1) = -\tan^2 x$

$\quad = \sec^2 x \cdot -\sin^2 x$

$\quad = \dfrac{1}{\cos^2 x} \cdot -\sin^2 x = -\tan^2 x$

8. $\dfrac{\cos x}{1-\sin x} = \sec x + \tan x$

$\quad = \dfrac{\cos x(1+\sin x)}{(1-\sin x)(1+\sin x)}$

$\quad = \dfrac{\cos x(1+\sin x)}{\cos^2 x}$

$\quad = \dfrac{(1+\sin x)}{\cos x} = \sec x + \tan x$

Lecture 5.3 Solving Trigonometric Equations

Solve the following trigonometric equations: (Isolate the trig function)

1. $2 - 4\cos x = 0$ Solutions in the interval: $[0, 2\pi)$

$\dfrac{\pi}{3}, \dfrac{5\pi}{3}$

All Solutions possible (General):

$\dfrac{\pi}{3} + 2\pi n, \dfrac{5\pi}{3} + 2\pi n, where\ n = integer.$

2. $\sin x + 2 = -3\sin x$ Solutions in the interval: $[0.2\pi)$

$\dfrac{7\pi}{6}, \dfrac{11\pi}{6}$

All Solutions possible (General):

$\dfrac{7\pi}{6} + 2\pi n, \dfrac{11\pi}{6} + 2\pi n$

3. Solve: $2\tan^2 x - 6 = 0$ in the interval: $[0.2\pi)$

$\dfrac{\pi}{3}, \dfrac{2\pi}{3}, \dfrac{4\pi}{3}, \dfrac{5\pi}{3}$

4. Solve: $\sec x \csc x = \csc x$ in the interval: $[0.2\pi)$

x= 0

5. solve: $2\cos^2 x + \cos x - 1 = 0$ in the interval: $[0.2\pi)$

$\dfrac{\pi}{3}, \pi, \dfrac{5\pi}{3}$

6. solve: $2\cos^2 x + 3\sin x - 3 = 0$ in the interval: $[0.2\pi)$

$\dfrac{\pi}{6}, \dfrac{\pi}{2}, \dfrac{5\pi}{6}$

7. solve: $\cos x + 1 = \sin x$ in the interval: $[0.2\pi)$

$\dfrac{\pi}{2}, \pi, \dfrac{3\pi}{2}$

8. solve: $2\sin 2t + 1 = 0$ in the interval: $[0.2\pi)$

$\dfrac{7\pi}{12}, \dfrac{11\pi}{12}, \dfrac{19\pi}{12}, \dfrac{23\pi}{12}$

9. solve: $2\cos 3\theta - 1 = 0$ in the interval: $[0.2\pi)$

$\dfrac{\pi}{9}, \dfrac{5\pi}{9}, \dfrac{7\pi}{9}, \dfrac{11\pi}{9}, \dfrac{13\pi}{9}, \dfrac{17\pi}{9}$

Lecture 5.4 Sum and Difference Formulas

1. Find the exact value of $\sin 75°$.

$$\frac{\sqrt{6} + \sqrt{2}}{4}$$

2. Find the exact value of $\cos \frac{7\pi}{12}$.

$$\frac{-\sqrt{6} + \sqrt{2}}{4}$$

3. Find the exact value of $\cos 58° \cos 13° + \sin 58° \sin 13°$.

$$\frac{\sqrt{2}}{2}$$

4. Find the exact value of $\tan 195°$

$$-\sqrt{3} + 2$$

5. Find the exact value of $\sin(x + y)$ given that $\sin x = \frac{3}{5}$, where $0 < x < \frac{\pi}{2}$ and $\cos y = \frac{-5}{13}$, where $\frac{\pi}{2} < y < \pi$.

$$\frac{33}{65}$$

6. If A is an angle in the third quadrant, B is an angle in the second quadrant, $\tan A = \frac{3}{4}$, and $\tan B = -\frac{1}{2}$, in which quadrant does angle $(A + B)$ lie?

Quadrant 1

7. Simplify each expression using the addition formulas:

a.) $\sin(90° - x) =$

$= \cos x$

b.) $\cos(180° + x) =$

$= -\cos x$

8. Find all solutions of $\sin\left(x + \frac{\pi}{4}\right) + \sin\left(x - \frac{\pi}{4}\right) = -1$, where $0 \leq x < 2\pi$.

$\dfrac{5\pi}{4}, \dfrac{7\pi}{4}$

Lecture 5.5 Double Angle Formulas

Develop the double angle formulas for sine and cosine.

$\sin(2u) =$ sin (u+u) = sin u cos u + cos u sin u = 2 sin u cos u

$\cos(2u) =$ cos(u+u) = cos u cos u – sin u sin u = cos²u -sin²u

1. Express $\sin 4x$ in terms of $\sin x$ and $\cos x$.

4sinx cos³x-4sin³xcosx

2. Solve the equation: $\sin 2x - \cos x = 0$ in the interval $[0, 2\pi)$

$$\frac{\pi}{2}, \frac{3\pi}{2}, \frac{\pi}{6}, \frac{5\pi}{6}$$

3. Find the value of $\cos 2\theta - \sin (90°+\theta)$ if $\tan \theta = -\frac{3}{4}$ and $\sin \theta$ is positive.

$$\frac{27}{25}$$

4. Solve the equation: $2 \cos x + \sin 2x = 0$ in the interval $[0, 2\pi)$

$$\frac{\pi}{2}, \frac{3\pi}{2}$$

5. Given $\sin x = \frac{12}{13}$ and $\frac{\pi}{2} < x < \pi$, find $\sin 2x, \cos 2x, and \tan 2x$.

$$\frac{-120}{169}, \frac{-119}{169}, \frac{120}{119}$$

6. Find the exact value of the $\cos 165°$ using the half-angle formula.

$$\frac{-\sqrt{2+\sqrt{3}}}{2}$$

7. Find the exact value of the $\sin 105°$ using the half-angle formula.

$$\frac{\sqrt{2+\sqrt{3}}}{4}$$

8. Prove: $\dfrac{\sin t+\sin 3t}{\cos t+\cos 3t} = \tan 2t$

$$= \frac{\sin t+\sin(t+2t)}{\cos t+\cos(t+2t)} = \frac{\sin t+\sin t\cos 2t+\sin 2t\cos t}{\cos t+\cos t\cos 2t-\sin t\sin 2t} = \frac{\sin t\,(1+\cos 2t+2\cos^2 t)}{\cos t(1+\cos 2t-2\sin^2 t)} = \frac{\sin t\,(4\cos^2 t)}{\cos t(\cos 2t+\cos 2t)}$$

$$= \frac{\sin t\,(4\cos^2 t)}{\cos t(\cos 2t+\cos 2t)} = \frac{\sin t\,(2\cos t)}{\cos 2t} = \frac{\sin 2t}{\cos 2t} = tan2t$$

9. Rewrite the product $\cos 5x\sin 4x$ as a sum or difference.

$$= \frac{1}{2}\left(\sin(9x) - \sin(x)\right)$$

10. Rewrite $\cos 4x + \cos 6x$ as a product.

$$= 2\cos(5x)\cos(-x)$$

11. Find the solutions of $\sin 5x + \sin 3x = 0$, where $0 \leq x < 2\pi$.

$$0, \frac{\pi}{4}, \frac{\pi}{2}, \pi, \frac{3\pi}{4}, \frac{5\pi}{4}, \frac{3\pi}{2}, \frac{7\pi}{4},$$

Lecture 6.1 Law of Sine

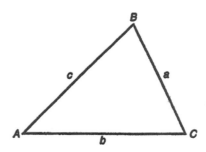

Given the following information, find the remaining sides and angles of the triangle.

1. $\angle B = 25°$, $\angle C = 38°$ and $b = 210$

A= 117°, a= 442.74, c= 305.92

2. $\angle A = 50°$, $\angle B = 28°$ and a= 21

b= 12.87, C= 102°, c= 26.81

Examine the height of the triangle (h). How can we relate that to $\angle A$?

h= c sinA°

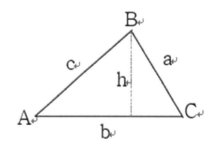

Given the following information, determine the number of triangles that exist.

3. $a = 14.8, b = 25.4$ and $\angle A = 82°$ Drawing

None.

4. a = 13, b = 21 and ∠A = 43°

<u>Drawing</u>

None.

5. a = 49, b = 35 and ∠B = 60°

<u>Drawing</u>

2 triangles

6. Find the area of $\triangle ABC$ if a = 180 inches, b = 150 inches, and ∠C = 30°

6750

7. Find the area of a triangular lot having two sides of 7.5 feet and 13 feet and the included angle of 120°.

42.22

Lecture 6.2 Law of Cosine

Solve the following triangles:

1. In triangle ABC, $a = 10, b = 6$, and $\angle C = 25°$.

c= 5.22

B= 29.07°

A= 125.93°

2. In triangle ABC, $a = 12, b = 8$, and $c = 6$.

a= 117.28°

b=3 6.34°

c= 26.38°

3. Find the area of a triangle with sides 13 inches, 15 inches, and 18 inches.

95.92

4. A surveyor finds that the edges of a triangular lot measure 42.5m, 37.0m, and 28.5m. What is the area of the lot?

121.86

Lecture 6.3 Basic Vector

1. Suppose we have vector $|\overrightarrow{PQ}|$, with $P = (1, 0)$ and $Q = (5, 3)$. Draw the vector and find the magnitude of the vector. (length of vector)

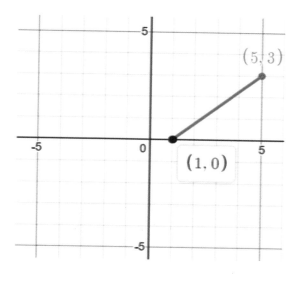

Magnitude $= 5$

Now, move the vector so that it starts at (0,0).
Where would the terminal point now be?

(4,3)

2. Find the component form of vector $\overrightarrow{ab}$, if a= $(-4, 9)$ and b= $(-5, -8)$.

(-1, -17)

3. What is $|\overrightarrow{ab}|$?

$\sqrt{290}$

4. Suppose $v = \langle 3, 7 \rangle$, Find $\|v\|$.

$\sqrt{58}$

5. Given the vectors u and v. Draw the following:

a.) u + v

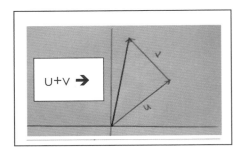

b.) 2u

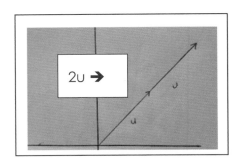

c.) u − v

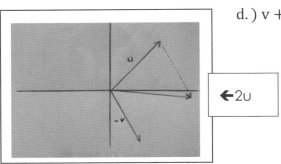

d.) v + .5u

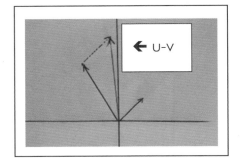

6. Let $v = \langle -6, 4 \rangle$ and $w = \langle 2, 5 \rangle$. Find the following algebraically:

a.) 2v

b.) w − v

c.) 2v + w

(-12,8)

(8,1)

(-10, 13)

7. Find a unit vector u in the direction of $v = \langle 3, -2 \rangle$.

$(3/\sqrt{13} , -2/\sqrt{13})$

8. Let u be the vector with initial point (3, -7) and terminal point (4,9). Write u as a linear combination of the standard unit vectors i and j.

$1i+16j$

9. Suppose u=-2i+5j and v=i-7j. Find 3u+4v.

$(-2i-13j) = (-2, -13)$

10. Find the direction angle of each vector. (Check quadrants)

a.) $u = 2i + 2j$

$45°$

b.) $u = -2i - 5j$

$248.20°$

11. Find the component form of the vector that represents the velocity of an airplane descending at a speed of 90 miles per hour at an angle of 210°.

$(-105\sqrt{3}, -105)$

12. Find the vector v with the given magnitude: $\|v\| = 12$ in the direction of $u = 2i - j$

$(24/\sqrt{5}, -12/\sqrt{5})$

Lecture 7.1 Sequences and Series

1. Find the first four terms of the following sequences:

a.) $a_n = 3n + 4$

7, 10, 13, 16

b.) $b_n = \dfrac{(-1)^n \cdot 2}{n^2 + 1}$

-1, $\dfrac{2}{5}$, , $-\dfrac{1}{5}$, $\dfrac{2}{17}$

2. Write an expression for the apparent n^{th} term of each sequence.

a.) $2, 6, 10, 14, 18, \ldots a_n$

$a_n = 4n-2$

b.) $3, 6, 11, 18, 27, \ldots a_n$

$a_n = n^2 + 2$

3. Write the first four terms of the recursively defined sequence:

a.) $a_1 = 2$, $a_{n+1} = 3a_n + 4$

2, 10, 34, 106

b.) $a_1 = -1$, $a_{n+1} = 2(a_n)^2 + 21$

-1, 23, 1079, 2328503

c.) $a_2 = -3$, $a_{n+1} = -a_n + 4$

-7, -3, +1, 5

4. Find the following factorials:

| $0! =$ | 1 | $1! =$ | 1 | $2! =$ | 2 | $3! = 6$ |
| $4! =$ | 24 | $5! =$ | 120 | $6! =$ | 720 | $7! = 5040$ |

5. Write the first four terms of the following sequence: $a_n = \dfrac{3n}{n!}$

(3, 3, 3/2, 1/2)

6. Evaluate the factorial expressions:

a.) $\dfrac{9!}{3!8!}$ $\dfrac{3}{2}$

b.) $\dfrac{2n!}{(n+1)!}$ $\dfrac{2}{(n+1)}$

c.) $\dfrac{2(n-2)!}{(n-3)!}$ 2(n-2)

d.) $\dfrac{(n+1)!}{(n+5)!}$ $\dfrac{1}{(n+5)(n+4)(n+3)(n+2)}$

7. Find the following sums: (expand and then simplify)

a.) $\sum_{i=1}^{5} 4i - 2 =$ 50

b.) $\sum_{i=2}^{7} (-1)^{i-1} \cdot i! =$ 4420

8. For the series $2 \cdot \sum_{i=1}^{\infty} \dfrac{3}{2^i}$, find (a) and (b)

a.) The third partial sum is:

$\dfrac{21}{4}$

b.) The approximate final sum would be:

6

9. Use sigma notation to write the sum.

a.) $1 + 4 + 7 + 10 + 13 + 16 + 19 + 22$

$$\sum_{n=0}^{7} 3n + 1$$

b.) $\frac{2}{1} + \frac{2}{2} + \frac{2}{6} + \frac{2}{24} + \frac{2}{120}$

$$\sum_{n=1}^{5} \frac{2}{n!}$$

10. Find the sum of $\sum_{k=0}^{4}(-2k + k!)$.

14

Lecture 7.2 Arithmetic Sequences and Partial Sums

1. Find the common difference in the following arithmetic sequences:

a.) $3, 8, 13, 15, \ldots$

5

b.) $25, 11, -3, -17, \ldots$

-14

c.) $1, \frac{2}{3}, \frac{1}{3}, 0, \frac{-1}{3}, \ldots$

$-\frac{1}{3}$

2. Determine if the sequence is arithmetic:

a.) $2, 6, 10, 4, 10, 12, \ldots$

no. d is not constant.

b.) $1, -4, -9, -14, \ldots$

yes. d= -5

c.) $1, \frac{1}{2}, \frac{1}{3}, \frac{1}{4}, \ldots$

no. d is not constant.

3. Find the 6th term of the arithmetic sequence with common difference 5 and first term 9.

$t_6 = 9+(5)5 = 34$

4. Find the nth term of the arithmetic sequence with fifth term 19 and ninth term 27.

$t_n = 11+(n-1)2$

5. Find the ninth term of the arithmetic sequence whose first two terms are 1 and 6.

$t_9 = 1+8 \times 5$

$= 41$

6. Find the sum: $1 + 6 + 11 + 16 + 21 + 26 + 31 + 36 + 41 + 46 + 51$ (Use the formula)

$= 286$

7. Find the sum of the first 12 multiples of 4.

$= 312$

8. Find the 15th partial sum of the sequence $2, 5, 8, 11, \ldots$

$= 345$

9. Find the sum: $\displaystyle\sum_{n=1}^{100} 2n =$

$= 10100$

10. Find the sum of the numbers from 1 to 1000.

$= 500500$

11. Find the sum: $\displaystyle\sum_{n=3}^{45} 3n + 5 =$

$= 3311$

Lecture 7.3 Geometric Sequences and Series

1. Find the common ratio of the following geometric sequence:

$2, 6, 18, 54$

$r = 3$

2. Determine whether the sequence is geometric. If it is, find the common ratio.

a.) $1, -2, 4, -8, \ldots$

yes, $r = -2$

geometric seq.

b.) $9, -6, 4, \dfrac{-8}{3}, \ldots$

yes, $r = -\dfrac{2}{3}$

c.) $\dfrac{1}{5}, \dfrac{2}{7}, \dfrac{3}{9}, \dfrac{4}{11}, \ldots$ no.

3. Write the first four terms of the geometric sequence in which $a_1 = 9$ and $r = \dfrac{-1}{3}$.

$9, -3, 1, -\dfrac{1}{3}$

4. Write the first four terms of the geometric sequence in which $a_1 = 243$ and

$a_{k+1} = \dfrac{1}{3} a_k.$

$243, 81, 27, 9$

5. Find the fifth term of a geometric sequence in which $a_1 = 3$ and $r = \dfrac{2}{3}$.

$3, 2, \dfrac{4}{3}, \dfrac{8}{9}, \dfrac{16}{27}$

6. Find the twenty-2nd term of the geometric sequence $1, 3, 9, 27, \ldots$

$t_n = 3^{21}$

7. Find the sum of $\sum_{n=2}^{11} 4(.2)^n$

$0.1999\ldots$

8. Find the sum of $\displaystyle\sum_{i=1}^{6} 64\left(\frac{1}{4}\right)^{i-1}$

$1365/16$

The Sum of an Infinite Geometric Series

Exploration:
Suppose we have a sequence in which $a_1 = 2$ and the sequence went on forever (infinite). Could we find a finite sum for the infinite series? Let's look at a few of cases:

Do the following series approach a specific value on your calculator?

a.) r = 1, $\quad 2 + 2 + 2 + 2 + 2 + \cdots$ No. it goes to infinity

b.) r = −1, $\quad 2 - 2 + 2 - 2 + 2 + \cdots$ No, it oscillates between 0 and 2.

c.) r = 2, $\quad 2 + 4 + 8 + 16 + 32 + \cdots$ No. it goes to infinity

d.) $r = \frac{1}{2}$, $\quad 2 + 1 + \frac{1}{2} + \frac{1}{4} + \frac{1}{8} + \cdots$ Yes, it is 4.

9. Evaluate $\displaystyle\sum_{n=1}^{\infty} 4(0.4)^{n-1}$

$\dfrac{20}{3}$

10. Find the total sum of $3 + 0.3 + 0.03 + 0.003 + 0.0003 + \ldots.$

3.333333... = 3+1/3 = $\dfrac{10}{3}$

11. Find the total sum of $\displaystyle\sum_{n=1}^{\infty} \frac{1}{2}(2)^n$

infinity

∞

Lecture 7.4 The Binomial Theorem

The Binomial Theorem is used to expand binomials to higher powers.

Let's examine powers of $(x + y)$

Draw Pascal's Triangle

$(x + y)^0 =$ 1

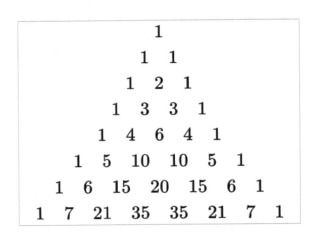

$(x + y)^1 =$ $x + y$

$(x + y)^2 =$ $x^2 + 2xy + y^2$

$(x + y)^3 =$ $x^3 + 3x^2y + 3xy^2 + y^3$

1. Find the binomial coefficients:

a.) $_7C_4$ b.) $_{12}C_0$ c.) $\binom{12}{3}$ d.) $\binom{6}{1}$

35   1   220   6

2. Using the fact from above, $_{17}C_2=$ Why is this true?

    $_{17}C_{15} = 136$ it cancels.

3. Use the Binomial Theorem to expand and simplify: $(x - 3y)^4$.

$x^4 - 12x^3y + 54x^2y^2 - 108xy^3 + 81y^4$

4. Use the Binomial Theorem to expand and simplify: $(2x - 3y)^3$

$= 8x^3-36x^2y+54xy^2-27y^3$

5. Expand the binomial using Pascal's triangle to determine the coefficients.

$(2a - b)^4$

$= 16a^4-32a^3b+24a^2b^2-8ab^3+b^4$

6. Find the coefficient a of the given term in the expansion of the binomial:

a.) $(2x^2 - 3)^{12}$, where the term is ax^{10}

-55427328

b.) $(2x + y)^{10}$, where the term is ax^2y^8.

180

7. Find the 4th term in the expansion of $(2x - 3y)^{12}$.

$-3041280x^9y^3$

8. Find the 3rd term in the expansion of $(3x + 4y)^7$.

$81648\ x^5y^2$

Lecture 8.1 Conic Section: Parabola

1. Graph $y = \frac{1}{4}x^2$. Label the vertex, focus, directrix, and axis of symmetry.

p= 1
vertex: (0,0)
focus: (0,1)
directrix: y= -1
Axis of sym.: x= 0

2. Graph $\left(x - \frac{1}{2}\right)^2 = 2(y - 3)$. Label the vertex, focus, directrix, and axis of symmetry.

p= 1/2
vertex= (1/2, 3)
focus = (1/2, 3.5)
directrix: y= 2.5
axis of sym.: x= 1/2

3. Graph the parabola $2y^2 - 4y - x + 5 = 0$ Label the vertex, focus, directrix, and axis of symmetry.

p= 1/8
vertex (3,1)
focus (25/8, 1)
directrix: x= 23/8
axis of sym.: y= 1

4. Graph the parabola: $y^2 - 6y - 4x + 17 = 0$. Label the vertex, focus, directrix, and axis of symmetry.

p= 1
vertex (2,3)
focus (3, 3)
directrix: x= 1
axis of sym.: y= 3

5. Graph the parabola: $4y^2 - 2x + 8 = 0$. Label the vertex, focus, directrix, and axis of symmetry.

p= 1/8
vertex (4,0)
focus (33/8, 1)
directrix: x= 31/8
axis of sym.: y= 0

6. Find the standard form of the equation of the parabola with Focus (1, -3) and Vertex (1, -2).

-4(y+2) = (x-1)²

7. Find the standard form of the equation of the parabola with Vertex (-3, 2) and directrix of x=1.

-16(x+3)= (y-2)²

Lecture 8.2 Ellipses

1. Find the center, vertices, covertices, and foci of the ellipse given by $\frac{x^2}{16} + \frac{y^2}{4} = 1$. Then Graph.

center $(0,0)$

$X_{str} = 4$

$Y_{str} = 2$

vertices: $(4, 0), (-4,0)$

covertices: $(0,2), (0,-2)$

foci: $(\sqrt{12},0), (-\sqrt{12},0)$

2. Sketch the graph of $\frac{(x-3)^2}{16} + \frac{(y+2)^2}{25} = 1$. Identify center, vertices, covertices, and foci.

center $(3, -2)$

vertices $(3,3)$ $(3, -7)$

covertices: $(-1, -2), (7, -2)$

foci $(3,1), (3,-5)$

3. Sketch the graph of $4x^2 + 9y^2 - 8x - 54y + 49 = 0$. Label center, vertices, covertices, and foci.

center $(1,3)$

vertices $(-2,3), (4,3)$

covertices $(1,5), (1,1)$

foci $(1+\sqrt{5},3)$ $(1-\sqrt{5},3)$

4. Find the center, vertices, foci, and eccentricity of the ellipse: $4x^2 + 9y^2 - 24x + 36y + 36 = 0$. Then, sketch its graph.

center $(3,-2)$

vertices $(0,-2),(6,-2)$

covertices $(3,0), (3,-4)$

foci $(3+\sqrt{5},-2), (3-\sqrt{5},-2)$

$e = \sqrt{5}/3$

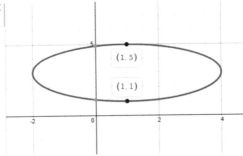

5. Find the standard form of the equation of the ellipse. Foci $(\pm 2, 1)$; Vertices $(\pm 5, 1)$

$x^2/25 + (y-1)^2/21 = 1$

6. Find the standard form of the equation of the ellipse. Foci $(0, \pm 2)$; Endpoints of the minor axis $(\pm 5, 0)$

$x^2/25 + y^2/29 = 1$

7. Find the standard form of the equation of the specified ellipse:
Foci: $(0,0), (6,0)$; Major axis of length 10.

$(x-3)^2/25 + y^2/16 = 1$

8. Find the standard form of the equation of the specified ellipse:
Vertices: $(\pm 4, 0)$; eccentricity of $\frac{1}{4}$.

$x^2/16 + y^2/12 = 1$

9. Find the standard form of the equation of the specified ellipse:
Endpoints of the major axis: $(\pm 8, 0)$; Endpoints of the minor axis $(0, \pm 4)$.

$x^2/64 + y^2/16 = 1$

Lecture 8.3 Extension (Identification of Conics)

Identify the graphs and then make a sketch. Label critical points.

1. $3y^2 - x + 1 = 0$ Parabola

2. $4x^2 - 9y^2 = 144$ Hyperbola

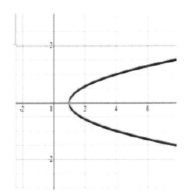

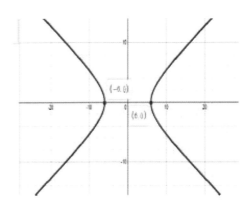

3. $x^2 + y^2 - 2y = 0$ Circle

4. $16x^2 + 9y^2 - 32x + 72y + 16 = 0$ Ellipse

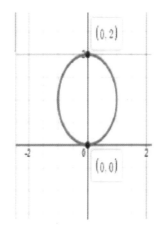

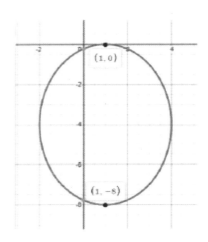

5. $-4x^2 + 25y^2 - 8x + 150y + 121 = 0$
   Hyperbola

6. $x^2 + y^2 + 6x - 8y = -9$
   Circle

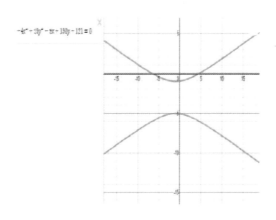

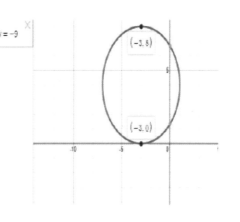

Lecture 8.4 Hyperbolas

1. Sketch the hyperbola: $\frac{(x-5)^2}{25} - \frac{(y+2)^2}{9} = 1$. Label the center, vertices, and foci.

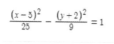

center (5,-2)
vertices (0,-2), (10.-2)
foci: $(5\pm\sqrt{34},-2)$

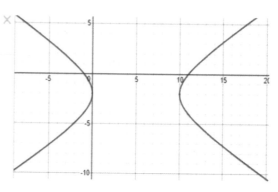

2. Sketch the hyperbola: $8y^2 - 50x^2 = 200$. Label the center, vertices, and foci.

center (0,0)
vertices (0,5), (0,-5)
foci: $(\pm\sqrt{29},0)$

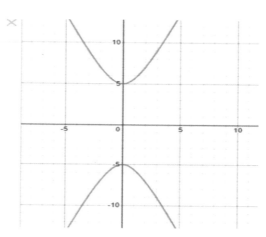

3. Given: $16y^2 - x^2 + 2x + 64y + 47 = 0$. Find the center, vertices, foci, and equations of the asymptotes. Sketch a graph.

center (1,-2)
vertices (1,-1), (1,-3)
foci: $(1, 2\pm\sqrt{17})$
eqn. of Asym.: (y+2)= ±1/4(x-1)
(y+2)²-(x-1)²/16 = 1

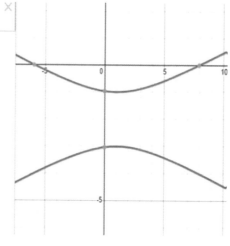

4. Find the standard form of the equation of the hyperbola with Vertices: $(\pm 3, 2)$ and Foci: $(\pm 6, 2)$

$- (y+2)^2/27 + x^2/9 = 1$

5. Find the standard form of the equation of the hyperbola with Vertices: $(0, \pm 3)$ and Asymptotes: $y = \pm 3x$.

$y^2/9 - x^2 = 1$

6. Find the standard form of the equation of the hyperbola that has Vertices: $(-2,1), (2,1)$ and passes through the point $(5,4)$.

$-7(y-1)^2/12 + x^2/4 = 1$

7. Find the standard form of the equation of the hyperbola that has Vertices: $(3,0)$, $(3, -6)$ and Asymptotes: $y = x - 6$ and $y = -x$.

$(y+3)^2/9 - (x-3)^2/9 = 1$

Lecture 8. 5 Parametric Equations

1. Sketch the graph of the curve given by $x = t+2$ and $y = t^2$ on $[-2, 5]$. Make a table of values and plot the points. Make sure to indicate the direction of the curve.

t	-2	-1	0	1	2	3	4	5
x	0	1	2	3	4	5	6	7
y	4	1	0	1	4	9	16	25

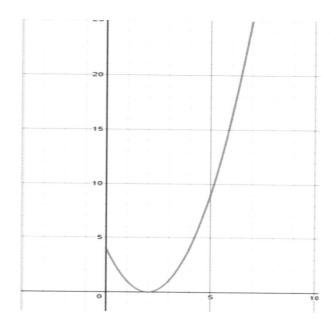

2. Use your graphing calculator to sketch the graph of the following curves on $[-2, 4]$.
   $x = 2t^2 \quad y = \dfrac{1}{2}t + 1$. Make sure to indicate the direction of the curve.

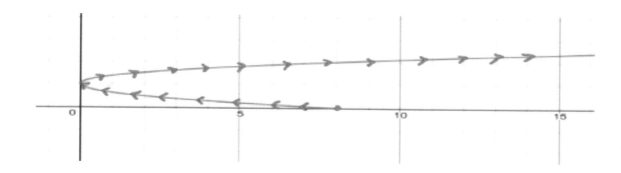

3. Use your graphing utility to graph the following parametric equation: $x = 2t^3$ and $y = t^2$. Is it a function?

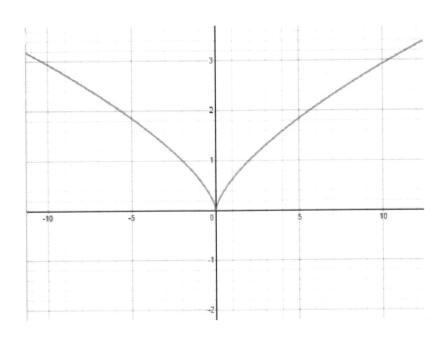

4. Eliminate the parameter of $x = 2t$ and $y = \frac{1}{2}t$. Also sketch the graph with the direction indicated.

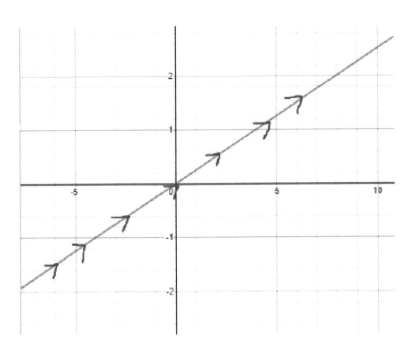

5. Eliminate the parameter of $x = 2\cos\theta$ and $y = 5\sin\theta$. Also sketch the graph with the direction.

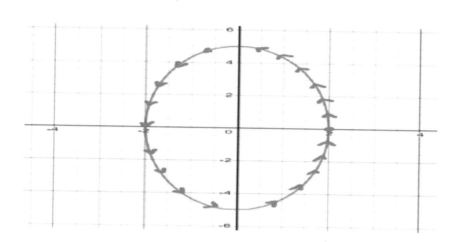

6. Now, find a set of parametric equations to represent the graph given by

$y = 2x^2 + 3$ using the following parameters.

a.) $t = x$

y= 2t²+3
x= t

b.) $t = x - 2$

x= t+2
y= 2(t+2)²+3
  = 2t²+8t+11

Lecture 8.6 Polar Coordinates

1. Plot the following points in the polar coordinate system.

a.) $\left(3, \frac{\pi}{4}\right)$

b.) $\left(4, \frac{-\pi}{6}\right)$

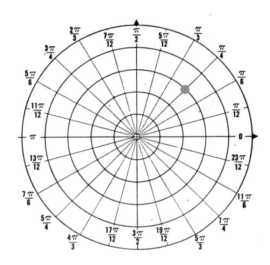

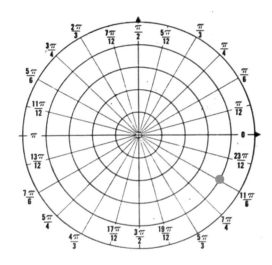

c.) $\left(-2, \frac{5\pi}{6}\right)$

d.) $\left(-4, -\frac{2\pi}{3}\right)$

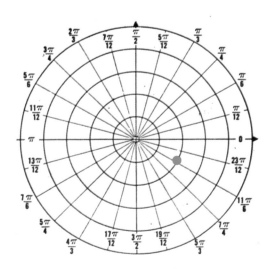

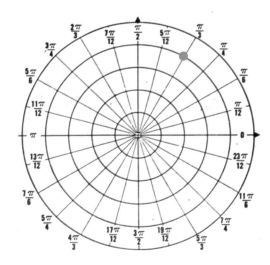

2. Plot the point $\left(1, \frac{\pi}{3}\right)$. Then find three other polar representations for the point.

<u>Three other representations:</u>

$(1, 7\pi/3), (-1, 4\pi/3), (-1, -2\pi/3)$

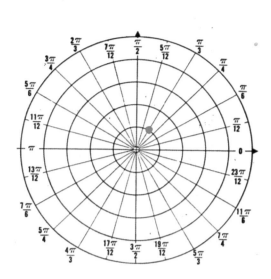

3. Convert the following polar coordinates to rectangular:

a.) $\left(6, \frac{\pi}{6}\right)$

$(3\sqrt{3}, 3)$

b.) $\left(-4, \frac{5\pi}{4}\right)$

$(2\sqrt{2}, 2\sqrt{2})$

4. Convert the following rectangular coordinates to polar coordinates:

a.) $(-2, -2)$

$(2\sqrt{2}, 5\pi/4)$

b.) $(0, -3)$

$(3, 3\pi/2)$

5. Convert the polar equations to rectangular form:

a.) $r = 6$

$x^2 + y^2 = 36$

b.) $\theta = \frac{\pi}{4}$

$y = x$

c.) $r = 2\cos\theta$

$(x-1)^2 + y^2 = 1$

d.) $r = \dfrac{2}{1+\sin\theta}$

$y = (-1/4) \cdot x^2 + 1$

6. Convert the rectangular equations to polar form:

a.) $x^2 + y^2 - 6x = 0$

r= 6 cosθ

b.) $3x - 6y + 2 = 0$

r= −2/(3cosθ-6sinθ)

c.) $y^2 = 2x$

r= 2cosθ/sin²θ

d.) $y = x$

θ= π/4

Lecture 8.7 Graphs of Polar Equations

1. Sketch the graph of $r = 2\sin\theta$. Fill in the table and plot the points. Use your graphing calculator to fill in the chart, then plot the points.

θ(degrees)	0	30°	60°	90°	120°	150°	180°	210°	270°	330°	360°
radians	0	$\frac{\pi}{6}$	$\frac{\pi}{3}$	$\frac{\pi}{2}$	$\frac{2\pi}{3}$	$\frac{5\pi}{6}$	π	$\frac{7\pi}{6}$	$\frac{3\pi}{2}$	$\frac{11\pi}{6}$	2π
r	0	1	$\sqrt{3}$	2	$\sqrt{3}$	1	0	-1	-2	-1	0

$r = 2\sin\theta$

$0 \ \underline{\quad} \ \leq \theta \leq \ \pi \ \underline{\quad}$

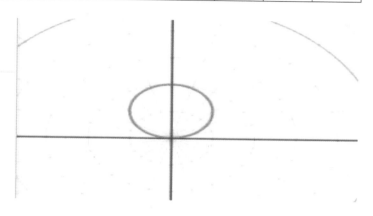

2. Determine the type of symmetry that the following polar graph has: $r = 2 - 2\sin\theta$. Then, sketch the graph. (you do not have to fill out the entire table if there is symmetry.)

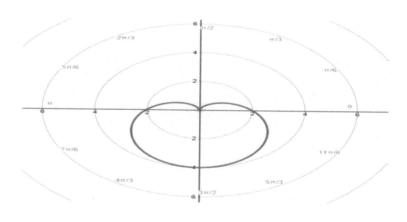

limacon graph

(degrees)	0	30°	60°	90°	120°	150°	180°	210°	270°	330°	360°
radians	0	$\frac{\pi}{6}$	$\frac{\pi}{3}$	$\frac{\pi}{2}$	$\frac{2\pi}{3}$	$\frac{5\pi}{6}$	π	$\frac{7\pi}{6}$	$\frac{3\pi}{2}$	$\frac{11\pi}{6}$	2π
r	2	1	.27	0	.27	1	2	3	4	3	2

3. Sketch the graph of $r = 2$

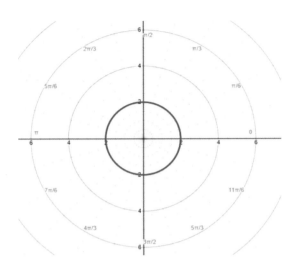

4. Sketch the graph of $\theta = \frac{\pi}{4}$.

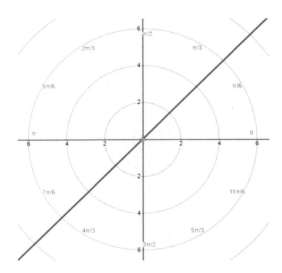

5. Sketch the graph of $r = 3 \sin 3\theta$. Identify the points where $|r|$ is maximum and points where $r = 0$. Use this to help you graph the polar graph.

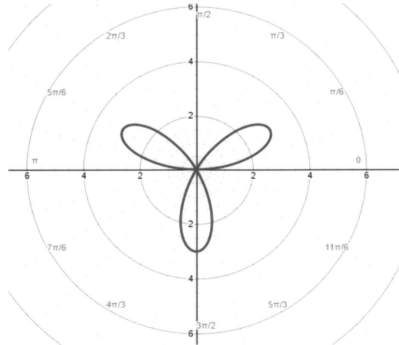

θ(degrees)	0	30°	60°	90°	120°	150°	180°	210°	270°	330°	360°
radians	0	$\frac{\pi}{6}$	$\frac{\pi}{3}$	$\frac{\pi}{2}$	$\frac{2\pi}{3}$	$\frac{5\pi}{6}$	π	$\frac{7\pi}{6}$	$\frac{3\pi}{2}$	$\frac{11\pi}{6}$	2π
r	0	3	0	-3	0	3	0	-3	3	-3	0

6. Sketch the graph of $r = 1 + 2\cos\theta$.

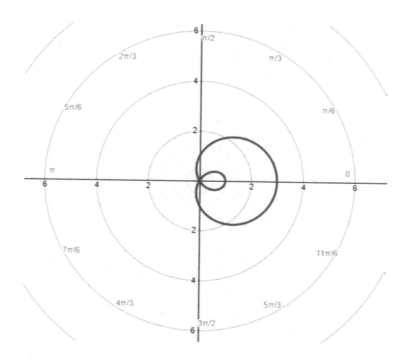

θ(degrees)	0	30°	60°	90°	120°	150°	180°	210°	270°	330°	360°
radians	0	$\dfrac{\pi}{6}$	$\dfrac{\pi}{3}$	$\dfrac{\pi}{2}$	$\dfrac{2\pi}{3}$	$\dfrac{5\pi}{6}$	π	$\dfrac{7\pi}{6}$	$\dfrac{3\pi}{2}$	$\dfrac{11\pi}{6}$	2π
r	3	2.7	2	1	0	-.7	-1	-.7	1	2.7	3

Lecture 9.1 Introduction to Limits

1. Use a table to evaluate the following limits with the aid of your calculator. (We must look at the function's value from both sides of the c)

$a.)\ \lim_{x\to3}(x-2)$

x	2.9	2.99	3	3.001	3.01	3.1
$f(x)$	0.9	.99	1	1.001	1.01	1.1

$b.)\ \lim_{x\to-1}\dfrac{x^2-1}{x+1}$

x	-1.1	-1.01	-1	-0.999	-0.99	-0.9
$f(x)$	-2.1	-2.01	2	-1.999	-1.99	-1.9

2. Use the graph of the function to evaluate the following limits. (Use your calculator.)

$a.)\ \lim_{x\to3}\dfrac{x-3}{\sqrt{x+3}-\sqrt{x}}$ 0

$b.)\ \lim_{x\to\pi}(\cos x)$ -1

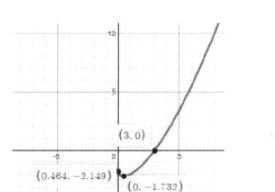

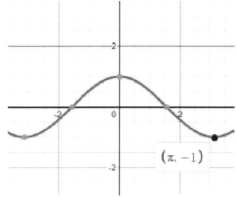

3. Do the following limits exist? (Why or why not)

a.) $\lim\limits_{x\to3}\dfrac{|x-3|}{x-3}$

No limit

left side and right

side limits are

different.

b.) $\lim\limits_{x\to0}\dfrac{2}{x}$

No limit

left side and right

side limits are

different.

c.) $\lim\limits_{x\to0}\sin\dfrac{1}{x}$

No limit because

sine oscillate

between -1 and 1,

there is no definite value

4. Evaluate the following limits using the properties and operations of limits.

a.) $\lim\limits_{x\to9}2\sqrt{x}$

6

b.) $\lim\limits_{x\to-1}-x^2-3x+7$

9

c.) $\lim\limits_{x\to0}\dfrac{3x+1}{x^2-3x-4}$

$-1/4$

d.) $\lim\limits_{x\to0}(12)$

12

e.) $\lim\limits_{x\to0}(2\sin x)$

0

f.) $\lim\limits_{x\to3}|x-3|$

0

5. Find the limit of $f(x)$ as x approaches 3., where $g(x) = \begin{cases} 7, x \neq 3 \\ -2, x = 3 \end{cases}$

No limit

Left-side and right-side limits are different.

6. Find the limit of $f(x)$ as x approaches 0., where $g(x) = \begin{cases} 3x - 7, x \geq 0 \\ -x^2 - 7, x < 0 \end{cases}$

-7

Lecture 9.2 Techniques for Evaluation Limits

1. Evaluate the limits:

$a.$) $\lim\limits_{x \to 1} 3x^2 + x - 5$

$b.$) $\lim\limits_{x \to 2} \dfrac{3x^2 - 5x - 1}{x^2 + 1}$

-1

$1/5$

2. Evaluate the limits:

$a.$) $\lim\limits_{x \to 1} \dfrac{x^2 - 1}{x - 1}$

$b.$) $\lim\limits_{x \to 2} \dfrac{x - 2}{x^3 - 8}$

2

$1/12$

3. Evaluate the limit: $\lim\limits_{x \to 0} \dfrac{\sqrt{x+2} - \sqrt{2}}{x}$

$\sqrt{2}/4$

A final technique is using your table feature and your graphing calculator.

4. Find the limit: $\lim\limits_{x \to 0} \dfrac{1 - \cos x}{x}$

0

5. Find the limit of $f(x)$ as x approaches 3.

a.) $f(x) = \dfrac{|x - 3|}{(x - 3)}$

b.) $f(x) = \begin{cases} x + 6, x \geq 3 \\ x^2, x < 3 \end{cases}$

No limit
Left-side and right-side limits are different.

9

6. For $f(x) = x^2 - 2x$, find $\lim\limits_{h \to 0} \dfrac{f(3+h) - f(3)}{h}$.

$4 + h$

Lecture 9.3 The Tangent Line Problem

1. Approximate the slope of the graph at (-1,1)   -2

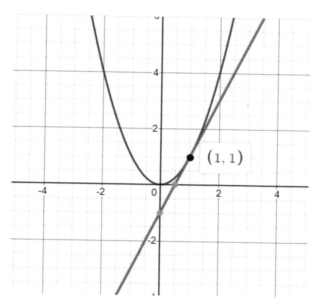

(1, 1)

2. Find the slope of the graph of $y = x^2$ at (1,1).

2

3. Find the slope of the graph of $f(x)=x^2 - 2$ at x.

2x

4. Find the derivative of the following function: $f(x)=x^2 + x - 3$

2x + 1

5. Find the derivative of the following function: $f(x)=\frac{1}{x}$. Then, find the slope of

the graph f at the points (1, 1) and $(-2,\frac{-1}{2})$

$m_1 = -1$

$m_{-2} = -1/4$

Lecture 9.4 Limits at Infinity and Limits of Sequences

1. Draw the graph of $f(x) = \frac{x}{x-2}$. Find the $\lim\limits_{x \to \infty} f(x)$ and $\lim\limits_{x \to -\infty} f(x)$.

1   1

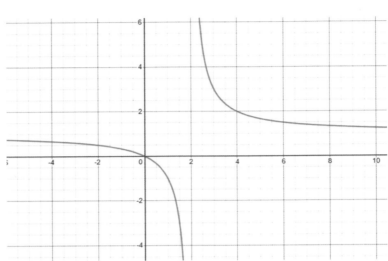

Evaluate the following limits:

2. $\lim\limits_{x \to \infty} \frac{2x^4 - 5}{x^4}$ 2

3. $\lim\limits_{x \to \infty} \frac{3x - 1}{x^2}$ 0

4. $\lim\limits_{x \to \infty} \frac{x^2 - 1}{5x^2}$ 1/5

5. $\lim\limits_{x \to \infty} \frac{2x^3 - 1}{7x^2 - x}$ ∞

6. Find the limit of the following sequences:

a.) $\lim\limits_{n\to\infty} \dfrac{3n+5}{2n^2+1}$ 0

b.) $\lim\limits_{n\to\infty} \dfrac{5n^2}{2n^2+1}$ 5/2

c.) $\lim\limits_{n\to\infty} \left[\dfrac{2}{n^4}\left(\dfrac{n^2(n+1)^2}{3}\right)\right]$ 2/3

WorkSheet

Answer Sheet

PreCalculus Worksheet

It is prepared so that you can learn the important parts
that you studied in PreCalculus Note once more
thoroughly.

1. $f = \{(2,3), (-3,4), (4,3)\}$

a) What is the domain of f?   2, 4, -3

b) What is the range of f?   3, 4

c) Is f a function?   yes

2. Which of the following equations represents y as a function of x? (Answer yes or no)

    a. $x^2 - 2y + 3 = 0$ b. $2x + y^2 - 5 = 0$

        yes   no

3. $f(x) = x^2 - 3x + 2$, find the following

    a. $f(2) =$ 0   b. $f(a) =$ $a^2 - 3a + 2$

    c. $f(x + h) =$ x²+2xh+h²-3x-3h+2

4. Find the domain of the following functions:

    a. $f(x) = x^3 + 3x - 2$ domain= all real number

    b. $f(x) = \frac{-2}{x+3}$ domain= x ≠-3

    c. $f(x) = \sqrt[4]{3 - x}$ domain= 3≥x

5. $f(x) = x^2 + 2x - 1, find \frac{f(x+h)-f(x)}{h} =$ 2x+h+2

6. f(x)= x³-5x+2,   find $\frac{f(x+h)-f(5)}{x+h-5} =$ (x³+3x²y+3xy²+y³-5x-5h-100)/(x+h-5)

PreCalculus Section 1.2 Work Sheet

Find the domain and range of the following functions:

1. $f(x) = x^3 - 4x + 3.$ Domain= all real number Range= all real number

2. $h(x) = -|x - 2|$ Domain= all real number Range= $y \leq 0$

Determine whether y is a function of x.

3. $2x - y^2 = 1$ no

4. $y = \frac{1}{5}x^5$ yes

Determine the intervals over which the function is increasing, decreasing, or constant and determine whether the function is even, odd, or neither.

5. $f(x) = x^2 - 4x$

   Increasing x>2 Decreasing x<2 Constant None

   Is it even or odd? Neither

6. $f(x) = -x^6 - 3x^4$

   Increasing x<0 Decreasing x>0 Constant None

   Is it even or odd? even

7. Graph the piecewise-defined function:

a.) $f(x) = \begin{cases} x^2 + 4, & x \le 1 \\ -x^2 + 4x + 3, & x > 1 \end{cases}$

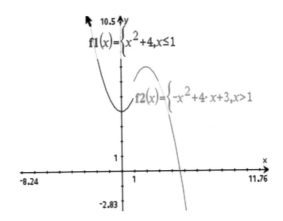

Is the function continuous?   no

b.) $f(x) = \begin{cases} x^2 - 4, & x \le 2 \\ \ln(x - 1), & x > 2 \end{cases}$

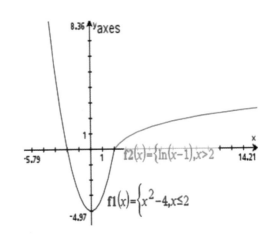

Is the function continuous?   yes

PreCalculus Section 1.3 Work Sheet

1. Sketch the graph of the three functions:

    $a.)\ f(x) = \sqrt{x}$ $b.)\ f(x) = \frac{1}{2}\sqrt{x}$ $c.)\ f(x) = -\frac{1}{2}\sqrt{x-3}$

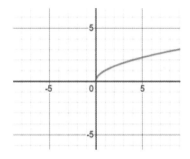  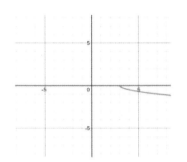

2. Write the equations for the following common functions: (asumme 1 interval = 1 unit.)

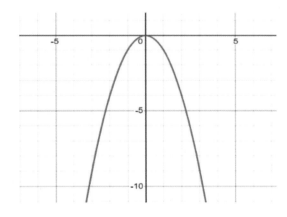 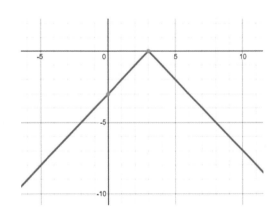

    a. $y = -x^2$ b. $y = -|x-3|$

3. $g(x) = 2(x-3)^2$

    a.) Identify the common function f that is related to g.

    b.) Describe the sequence of transformations from f to g.

    c.) Sketch the graph of g

       many possible answers..!!

    a.) $f(x) = 2x^2$

    b.) right 3 units   c.)

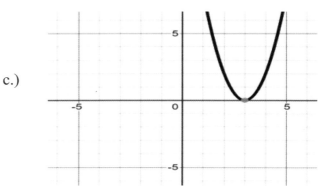

4. $h(x) = -(-x+1)^2 - 3$

    a.) Identify the common function f that is related to h.

    b.) Describe the sequence of transformations from f to h.

    c.) Sketch the graph of h

    a.) $f(x) = (x+1)^2 - 3$ c.)

    b.) reflecting over x and y-axis

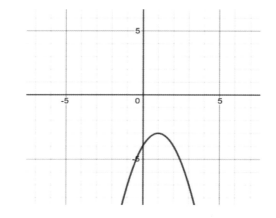

PreCalculus Section 1.4 Work Sheet

1. $f(x) = 3x - 5$ and $g(x) = 2 - x$. Find the following:

a.) $(f + g)(x) =$ $2x - 3$

b.) $(f - g)(x) =$ $4x - 7$

c.) $(fg)(x) =$ $-3x^2 + 8x - 10$

d.) $\left(\frac{f}{g}\right)(x) =$ $(3x-5)/(2-x)$

e.) what is the domain of $\frac{f}{g}$?   All except x= 2

f.) $(fg)(2) =$ 0

g.) $(f - g)(-5) =$ -27

2. $f(x) = |x + 1|$, and $g(x) = x - 4$, find the following:

a.) $f \circ g =$ $|x - 3|$

b.) $g \circ f =$ $|x + 1| - 4$

3. Find two functions f and g such that $(f \circ g)(x) = h(x)$.

a.) $h(x) = \frac{4}{(5x+2)^2}$ $f(x) = $ $4/x^2$

                                    $g(x) = $ $(5x + 2)$

PreCalculus Section 1.5 Work Sheet

1. Determine whether or not the function is one-to-one. (yes or no)

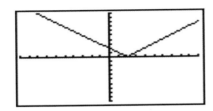

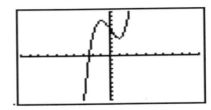

a.)　no

b.)　no

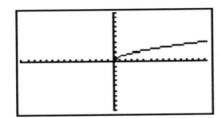

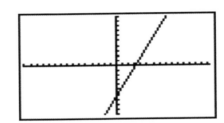

c.)　yes

d.)　yes

2. Find the inverse of $f(x) = \frac{2x-1}{3}$ algebraically. Also Verify your results showing that $f(f^{-1}(x)) = x$ and $f^{-1}(f(x)) = x$.

Verification　　　　　　　　　　　**Inverse=**　y^{-1}= (3x+1)/2

(2((3x+1)/2)-1)/3

= 3x/3= x

3. Determine algebraically whether the function is one-to-one.

$f(x) = \sqrt{x-3}$　　　　　Is $f(x)$ one-to-one?　　　　yes

4. Are $f(x) = \sqrt[3]{2x-10}$ and $g(x) = \frac{x^3+10}{2}$ inverses of each other?　　yes

5. Sketch the graph of the inverse functions on the following graphs:

a.)

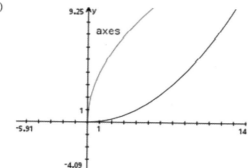

b.)

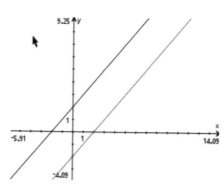

PreCalculus Section 2.1 Work Sheet

1. What is the degree of the following functions?

a.) $h(x) = 2 - 4x^4 + x^3$ Degree=   4

b.) $y(x) = -3x + 2x^3$ Degree=   3

2. Sketch the following functions: (identify vertex, intercepts, and axis of symmetry)

a.) $f(x) = 2x^2 - x + 1$

vertex (0.25, 0.875)

x-intercept: None

y-intercept: 1

Axis of Sym.: x= 0.25

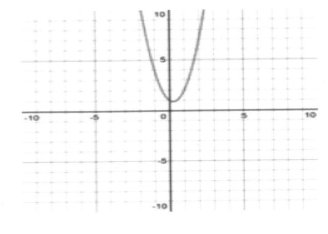

b.) $g(x) = -3x^2 + 9x + 1$

vertex (1.5, 7.75)

x-intercept: -0.107 and 3.107

y-intercept: 1

Axis of Sym.: x= 1.5

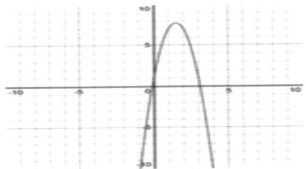

3. Find the quadratic function that has the indicated vertex and whose graph passes through the given point. Vertex: (4,-1) Point (2,3)

y= (x-4)²-1

4. Find the quadratic function that has the indicated vertex and whose graph passes through the given point. Vertex: (2,3) Point (0,2)

y= $\frac{-1}{4}$(x-2)²+3

5. Determine the coordinates of the vertex and the equation of the axis of symmetry of $y = 3x^2 + 2x - 5$. Does the quadratic function have a minimum or maximum value? If so, what is it?

vertex $= (-\frac{1}{3}, -\frac{16}{3})$ Axis of sym.: $x = -\frac{1}{3}$

min. $= -\frac{16}{3}$

PreCalculus Section 2.2 Work Sheet

1. Match the polynomial function with its graph:

   a.) $y = -2x + 3$

   b.) $y = -2x^2 + 2x + 1$

   c.) $y = x^4 + 3x + 1$

   d.) $y = \frac{1}{5}x^5 - 2x^3 + \frac{9}{5}x$

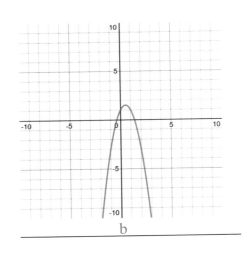

b

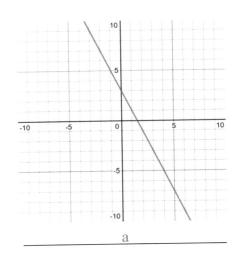

a

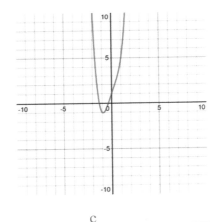

c

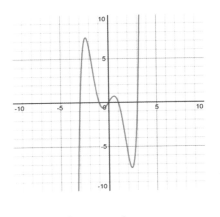

d

2. Use the Leading Coefficient Test to determine the right-hand and left-hand behavior of the graph of the polynomial function.

   a.) $f(x) = \frac{2}{3}x^3 - 3x$ right: up left: down

   b.) $h(x) = 2 - 3x^6$ both down

3. Find all the real zeros of the polynomial function.

a.) $f(x) = 81 - x^2$

-9, 9

b.) $(x) = x^4 - 6x^3 - 7x^2$

-1, 0, 7

4. Graph the following function: $f(x) = -4x^3 + 4x^2 + 15x$

y-int:(0,0)

x-int: (-1.5,0),(0,0),(2.5,0)

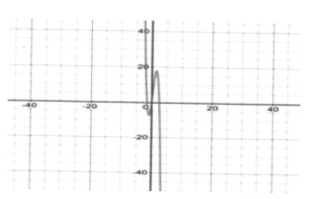

5. Find a polynomial function that has the given zeros.

a.) -5, -1, 0, 1, 2

y= (x+5)(x+1)(x-0)(x-1)(x-2)

PreCalculus Section 2.3 Work Sheet

1. Divide $x^3 - 9$ by $x^2 + 1$ using long division.

$x - \dfrac{(x+9)}{x^2+1}$

2. Divide $5x^3 + 6x + 8$ by $x + 2$ using synthetic division.

$5x^2 - 10x + 26 - \dfrac{44}{(x+2)}$

3. What is the remainder when $3x^3 + 2x^2 - 5x - 8$ is divided by $x + 2$?

-14

4. If $3+2i$, 2, and $2-3i$ are all zeros of $P(x) = 3x^5 - 36x^4 + 2x^3 - 8x^2 + 9x - 338$, what are the other zeros?

3-2i and 2 + 3i

5. Use synthetic division to show that x is a solution of the third-degree polynomial equation and use the result to factor the polynomial completely. List all the real z eros of the function. $x^3 - 28x - 48 = 0$, $x = -4$ is a solution.

    Complete Factorization   (x+4)(x-6)(x+2)

    All real Zeros   - 4, -2, 6

6. Use the Rational Zero Test to list the possible rational roots and then find all of the rational roots. $f(x) = x^3 - 4x^2 - 4x + 16$

    Possible Rational Roots ±1, ±2 ,±4, ±8, ±16

    Rational Roots - 2, 2, 4

7. Find all the real solutions of $x^4 - 13x^2 - 12x = 0$

    Solutions   x= 0, -3, -1, 4

PreCalculus Section 2.4 Work Sheet

1. Solve for a and b; $(a + 6) + 2bi = 4 - 5i$ a= -2 b= $-\frac{5}{2}$

2. Write in standard form: $-\sqrt{-75} + 3$ $3 - 5i\sqrt{3}$

3. Simplify and write in standard form:

   a.) $(13 - 2i) + (-3 + 6i) =$ $10 + 4i$

   b.) $(6 - 2i)(2 - 3i) =$ $6 - 22i$

   c.) $(3 + \sqrt{-5})(7 - \sqrt{-10}) =$ $21 - 3i\sqrt{10} + 7i\sqrt{5} + 5\sqrt{2}$

4. Find the product of the number and its conjugate:

   a.) $(7 - 12i)$ $-95 - 168i$

5. Simplify and write in standard form:

   a.) $\left(\frac{5}{1-i}\right)$ $\frac{5}{2} + \frac{5}{2}i$

   b.) $\left(\frac{8-7i}{1-2i}\right)$ $\frac{22}{5} + \frac{9}{5}i$

6. Simplify the complex number and write in standard form:

   a.) $i^2 + i^{23} - 7i$ $-1 - 8i$

   b.) $\frac{3}{(2i)^3}$ $\frac{3}{8}i$

7. Plot and label the complex numbers in the complex plane:

    a.) 2+3i   b.) -3i+5   c.) 5i   d.) -3   e.) 5-3i

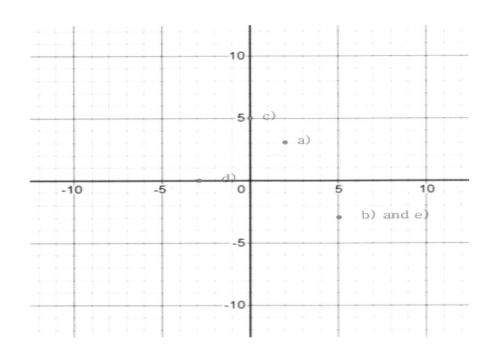

PreCalculus Section 2.5 Work Sheet

1. Find all the zeros of the function and write the polynomial as a product of linear factors.

   a.) $g(x) = x^2 + 10x + 23$

   Zeros $-5 \pm \sqrt{2}$

   Factors $(x-(-5+\sqrt{2}), (x-(-5-\sqrt{2}))$ or $(x+5-\sqrt{2}), (x+5+\sqrt{2})$

   b.) $h(x) = x^3 - 3x^2 + 4x - 2$

   Zeros $1, 1 \pm i$

   Factors $(x-1) \cdot (x-(1+i)) \cdot (x-(1-i))$

2. Find a polynomial function with integer coefficients that has the given zeros: 4, 3i, -3i

   Function $f(x) = x^3 - 4x^2 + 9x - 36$

3. Given: $f(x) = x^4 + 6x^2 - 27$

   a.) write $f(x)$ as the product of factors irreducible over the rationals:

   $(x^2+9) \cdot (x^2-3)$

   b.) write $f(x)$ as the product of linear and quadratic factors that are irreducible over the reals:

   $(x^2+9) \cdot (x+\sqrt{3}) \cdot (x-\sqrt{3})$

   c.) write $f(x)$ in completely factored form:

   $(x+3i) \cdot (x-3i) \cdot (x+\sqrt{3}) \cdot (x-\sqrt{3})$

4. Find all the zeros of $f(x) = 2x^4 - x^3 + 7x^2 - 4x - 4$ given 2i is a root.

   Zeros $2i, -2i, -1/2, 1$

   Complete Factorization $2(x-2i)(x+2i)(x+1/2)(1x-1)$ or $(x-2i)(x+2i)(2x+1)(x-1)$

PreCalculus Section 2.6 Work Sheet

1. Find the domain and the asymptotes of the following functions:

   a.) $f(x) = \frac{x+5}{x^2+2x-3}$ Domain: $x \neq -3, 1$ Horizontal Asymptote: $y = 0$

                                                 Vertical Asymptote: $x = -3$, $x = 1$

   b.) $f(x) = \frac{x^2-8}{2x-9}$ Domain: $x \neq 9/2$ Vertical Asymptote: $x = \frac{9}{2}$

                                                 Slant Asymptote: $y = \frac{1}{2}x + \frac{9}{4}$

2. Given: $f(x) = \frac{x}{2x-5}$ find the following and then graph:

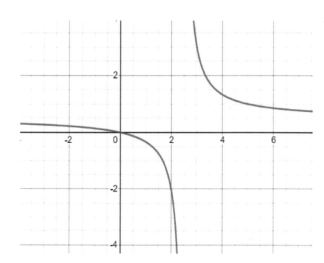

       Domain: $x \neq \frac{5}{2}$

       Vertical Asymptote: $x = \frac{5}{2}$

       Horizontal Asymptote: $y = \frac{1}{2}$

3. Given: $f(x) = \frac{x^2}{x^3-8}$

       Domain: $x \neq 2$

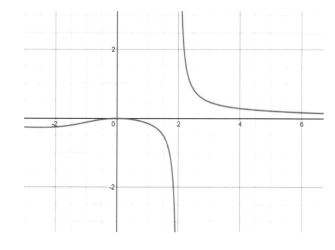

       Vertical Asymptote: $x = 2$

       Horizontal Asymptote: $y = 0$

PreCalculus Section 2.7 Work Sheet

1. Sketch the graph of $f(x) = \frac{1-2x}{x}$

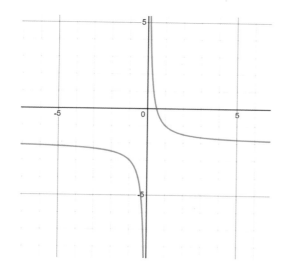

x-intercepts: $(\frac{1}{2}, 0)$

y-intercepts: none

verticalasymptotes: $x = 0$

horizontal asymptotes: $y = -2$

slantasymptotes: none

Symmetry: none

Additional Points

x	-2	-1	1	2	3
f(x)	-5/2	-3	-1	-3/2	-5/3

2. Sketch the graph of $g(x) = \frac{x^3}{2x^2-8}$

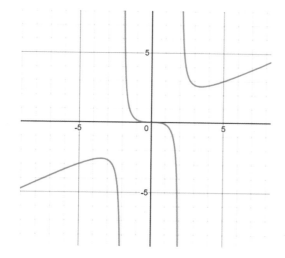

x intercepts: $(0,0)$

y intercepts: $(0,0)$

vertical asymptotes: $x = \pm 2$

horizontal asymptotes: none

slant asymptotes: $y = \frac{1}{2}x$

Symmetry: origin

Additional Points

x	-3	-1	0	4	5
f(x)	-2.7	1/6	0	8/3	2.976

PreCalculus Section 3.1 Work Sheet

1. Use your calculator to evaluate $e^{2.58}$ to three decimal places: 13.197

2. Graph and label the following functions on the same graph: (label intercepts and asymptotes)

   a.) $f(x) = \left(\frac{5}{2}\right)^x$ 　　　　　　b.) $g(x) = \left(\frac{5}{2}\right)^{-x}$ 　　　　　　c.) $h(x) = \left(\frac{5}{2}\right)^{x+2}$

   d.) $j(x) = -\left(\frac{5}{2}\right)^{x+2} - 1$

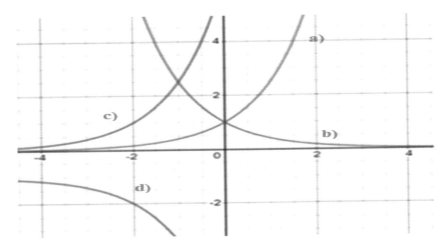

3. Use a graphing calculator to construct a table of values, then sketch the graph of

   $f(x) = e^{-x}$

x	f(x)
-3	20.1
-2	7.39
-1	2.72
0	1
1	0.37
2	0.14
3	0.05

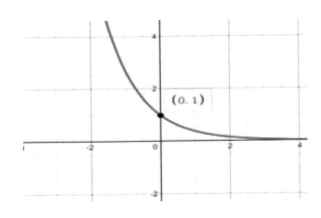

4. Complete the table to determine the balance A for $12,000 invested at a rate r=6% compounded continuously for t years.

t	1	10	20	30	40	50
A	12742.04	21865.43	39841.40	72595.77	132278.12	241026.44

5. The population of a town increases according to the model $P(t) = 2500e^{.0293t}$ where t is the time in years, with t=0 corresponding to 1990.

a.) Find the population in 1992, 1995, and 1998.

1992= 2651   1995= 2894   1998= 3160

1. Write the logarithmic equation in exponential form

   a.) $\log_3 27 = 3$ $3^3 = 27$

   b.) $\log_{16} 2 = \frac{1}{4}$ $16^{(1/4)} = 2$

2. Write the exponential equation in logarithmic form.

   a.) $5^3 = 125$ $\log_5 125 = 3$

   b.) $e^x = 3$ $x = \ln 3$

3. Evaluate the expression without a calculator.

   a.) $\log_{27} 9$ $\frac{2}{3}$

   b.) $\log_2\left(\frac{1}{8}\right)$ -3

4. Solve the equation for x.

   a.) $\log_7 7 = x$ 1

   b.) $\log_3 3^{-1} = x$ -1

5. Use a calculator to evaluate the logarithm. Round to three decimal places.

   a.) $\log_{10} 65$ 1.813

   b.) $-8.5 \ln 14$ -22.432

6. Find the domain, vertical asymptote, x intercept and sketch its graph by hand. Verify using your graphing calculator.

   a.) $g(x) = \log_{10} x$

   Domain : $(0,\infty)$

   Vertical asymptote : $x=0$

   x-intercept : $(1,0)$

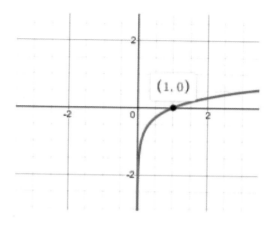

   b.) $f(x) = -\log_{10}(x + 3)$

   Domain : $(-3,\infty)$

   Vertical asymptote : $x = -3$

   x-intercept : $(-2, 0)$

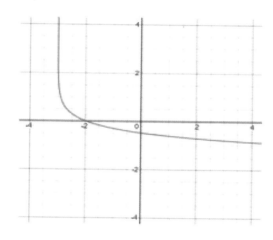

PreCalculus Section 3.3 Work Sheet

1. Evaluate the logarithm using the change-of-base formula. Round your result to three decimal places.

   a.) $\log_3 42$ 3.402

   b.) $\log_{\frac{1}{6}} 36$ -2

   c.) $\log_{\frac{1}{3}}(615)$ -5.845

2. Rewrite the logarithm as a multiple of (a) a common logarithm and (b) a natural logarithm.

   $\log_x \dfrac{3}{4}$ a.) $\log\left(\frac{3}{4}\right) / \log(x)$ b.) $\ln\left(\frac{3}{4}\right)/\ln x$

3. Use the properties of logarithms to write the expression as a sum, difference, and / or constant multiple of logarithms. (Assume all variables are positive)

   a.) $\log_{10}(10a)$ $1 + \log a$

   b.) $\log_6 b^{-3}$ $-3\log_6 b$

   c.) $\ln\left(\frac{x^3-1}{2x^3}\right), x > 1$ $\ln(x^3-1) - \ln 2 - 3\ln x$

   d.) $\ln\sqrt{x^2(x + 2)}$ $\ln|x| + \frac{1}{2}\ln(x+2)$

4. Write the expression as the logarithm of a single quantity.

   a.) $\ln a + 2\ln b$ $\ln(ab^2)$

   b.) $\frac{7}{2}\log_e(c - 4)$ $\ln\sqrt{(c-4)^7}$

   c.) $2\ln 6 + 5\ln d$ $\ln(36d^5)$

   d.) $2[\ln x - \ln(x + 1) - \ln(x - 1)]$ $\ln(x/(x^2-1))^2$

5. Find the exact value of the logarithm without using a calculator, if possible. If not possible, state why.

a.) $\log_5 \sqrt[3]{5}$ $\dfrac{1}{3}$

b.) $\log_4 4 + \log_4 16$ 3

c.) $\log_4(-64)$ no solution

PreCalculus Section 3.4 Work Sheet

1. Use a graphing utility to graph f and g in the same viewing window. Approximate the point of intersection of the graphs. Then solve the equation $f(x) = g(x)$ algebraically.

   a.) $f(x) = 9 \quad g(x) = 27^x$

   Approximation: 2/3

   Algebraic Solution: 2/3

   b.) $f(x) = 3\log_5 x \quad g(x) = 6$

   Approximation: 25

   Algebraic Solution: 25

2. Solve for x.

   a.) $3^x = 81$ 4

   b.) $\left(\frac{3}{4}\right)^x = \frac{27}{64}$ 3

   c.) $\ln(2x + 5) = 7$ (e⁷-5)/2

3. Simplify the expression

   a.) $\ln e^{3x-1}$ 3x - 1

   b.) $e^{3\ln x}$ x³

4. Solve the exponential equation algebraically. Round your result to three decimal places. Use your graphing calculator to verify your results.

   a.) $6^{5x} = 3000$ 0.89369

   b.) $1000e^{-4x} = 75$

         0.647567

   c.) $\frac{525}{1+e^{-x}} = \frac{275}{1}$

         0.09531

5. Use your graphing calculator to find an approximate solution to three decimal places.

   a.) $4^{\frac{-x}{2}} = 0.10$ 3.322

   b.) $\frac{119}{e^{6x}-14} = 7$ 0.572

6. Solve the logarithmic equation algebraically. Round your answer to three decimal places.

   a.) $\ln 4x = 1$

   $\frac{e}{4}$

PreCalculus Section 3.5 Work Sheet

1. Sketch a scatter plot of the data, decide whether the data could be modeled by a linear, exponential, or a logarithmic model. calc. needed.!

   a.) (1,11), (1.5,9.6), (2,8.2), (4,4.5), (6,2.5), (8,1.4)

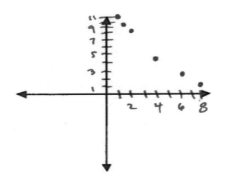

   Linear y=-1.376x+11.361,r=-.970

   Exponential: y=14.819(.744)x, r=-.99996

   Log: y=11.297-4.811 ln   x, r=-.998

   Exponential → b/c it has the best r

2. Complete the table for the time t necessary for P dollars to triple if interest is compounded annually at rate r.

r	2%	4%	6%	8%	10%	12%
t	55.48	28.01	18.85	14.27	11.53	9.69

3. The population P of a city is $P = 240{,}360e^{0.012t}$, where t=0 represents 2000. According to this model, when will the population reach 275,000?

11.219 years later so during 2011

PreCalculus Section 4.1 Work Sheet

1. Convert each degree measure to radian.

   a.) $150°$ $5\pi/6$

   b.) $-225°$ $(-5\pi)/4$

2. Convert each radian measure to degree measure.

   a.) $\frac{5\pi}{3}$ $300°$

   b.) $\frac{-12\pi}{6}$ $-360°$

3. Determine the quadrant in which the angle lies.

   a.) $\frac{55\pi}{3}$ $q1$

   b.) $\frac{2.35}{2\pi}$ $q2$

   c.) $\frac{-5\pi}{6}$ $q3$

   d.) $\frac{-35\pi}{4}$ $q3$

4. Find the complement and supplement of $\frac{2\pi}{7}$.

   complement angle $= 3\pi/14$

   supplement angle $= 5\pi/7$

5. Determine the length of the arc of a circle of radius 5.25ft intercepted by a central angle having a measure of 57^0.

   5.223ft

1. Give the exact value (if defined) of the six trigonometric functions of $\theta = \frac{-5\pi}{6}$.

$\sin\theta = -1/2$ $\csc\theta = -2$

$\cos\theta = -\sqrt{3}/2$ $\sec\theta = -2/(\sqrt{3}) = (-2\sqrt{3})/3$

$\tan\theta = \sqrt{3}/3$ $\cot\theta = \sqrt{3}$

2. Find the exact value of the $\tan\frac{17\pi}{6}$.

$(-\sqrt{3})/3$

3. Use the function value $\tan t = 3$ to find the values of each of the following:

     a.) $\tan(-t) =$ b.) $\cot(t) =$

         -3   1/3

4. Use a calculator to evaluate the following: (Round to four decimal places)

     a.) $\csc(7.89) =$ b.) $\tan(-120.4°) =$

         1.001   1.7045

PreCalculus Section 4.3 Work Sheet

1. Given: $\cos\theta = -\frac{2}{7}$, and θ is in quadrant III. Find the other five trigonometric functions by drawing a triangle.

$\sin\theta = \frac{-3\sqrt{5}}{7}$ $\csc\theta = \frac{7}{-3\sqrt{5}}$

$\cos\theta = -2/7$ $\sec\theta = -7/2$

$\tan\theta = \frac{\sqrt{45}}{2}$ $\cot\theta = \frac{2}{\sqrt{45}}$

2. Suppose $\cot\theta = 2$ and θ is in the third quadrant. Find the $\csc\theta$ by using a trigonometric identity.

$-\sqrt{5}$

3. Using a calculator, find the following to four decimal places:

   a.) $\csc 32.7°$ b.) $\tan 2.45$

      1.568 -0.828

   c.) $\sec 132°$ d.) $\cos -3.53$

      -1.4945 0.9981

4. A 23-foot ladder, leaning against the side of a house, reaches 12 feet up from the bottom corner of the house. What angle does the ladder make with the ground?

27.55°

PreCalculus Section 4.4 Work Sheet

1. State the quadrant in which θ lies.

   a.) $\sec\theta > 0 \ and \ \csc\theta < 0.$ 4th

   b.) $\cot\theta < 0 \ and \ \cos\theta < 0.$ 2nd

2. Find the values of the remaining five trigonometric functions if $\sec x = \frac{13}{-5}$ and $\cot x < 0.$

   $\sin\theta = 12/13$ $\cos\theta = -5/13$ $\tan\theta = -12/5$

   $\csc\theta = 13/12$ $\sec\theta = -13/5$ $\cot\theta = -5/12$

3. Find the reference angle for each of the following:

   a.) $\theta = 356°$ Reference angle= $4°$

   b.) $\theta = -453°$ Reference angle= $87°$

   c.) $\theta = \frac{15\pi}{3}$ Reference angle= $0°$

   d.) $\theta = \frac{-7\pi}{3}$ Reference angle= $\pi/3$

4. Using a calculator, evaluate each of the following to four decimal places.

   a.) $\tan 572.7° = 0.642$ b.) $\sec 160° = -1.0642$

5. Using a calculator, find two values of θ , where $0° \le \theta < 360°$. (Round values to one decimal place.)

   a.) $\sin\theta = .4226$ b.) $\cos\theta = -.6018$

      25°, 155°   127°, 217°

PreCalculus Section 4.5 Work Sheet

1. Find the amplitude, period, the phase shift and vertical displacement (if they exist).

   a.) $y = 3\cos\frac{x}{2} + 2$

   Amp.: 3

   Period: 4π

   no shift

   Vert. disp. +2

   b) $y = \frac{1}{2}\sin(3x - \pi)$

   Amp.: 1/2

   Period: $2\pi/3$

   P. shift: $\pi/3$

   Vert. disp.: None

   c) $y = -5\sin(2x - \frac{\pi}{2})$

   Amp.: 5

   Period: π

   P. shift: $\pi/4$

   Vert. disp.: None

2. Describe how the graph of the first equation can be obtained from the graph of the second equation.

   a.) $y = \sin(2x - 2\pi)$

   $y = \sin 2x$

   π unit to the right.

   b) $y = -3 + 4\cos(x - \frac{\pi}{4})$

   $y = 2\cos(x - \frac{\pi}{4})$

   increase Amp. by 2 and down 3 unit.

3. Graph each equation, showing one complete cycle of the curve.

   a.) $y = 3\cos 2x$

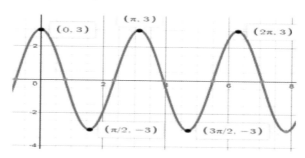

   b) $y = 2\sin 2(x + \frac{\pi}{4}) - 4$

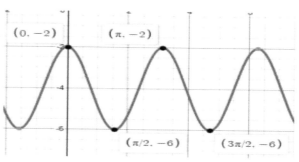

PreCalculus Section 4.6 Work Sheet

1. Graph each function, showing one complete cycle.

a.) $y = \tan\left(\frac{1}{2}x\right) + 2$

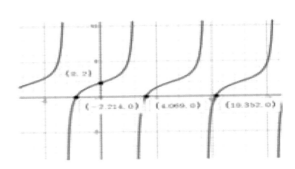

b.) $y = -3 + \frac{1}{2}\sec\left(2(x - \frac{\pi}{2})\right)$

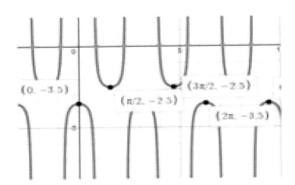

c.) $y = 3\csc(2\pi x) - 5$

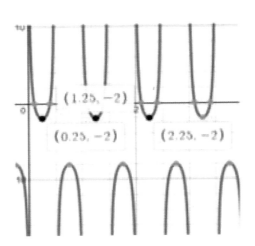

PreCalculus Section 4.7 Work Sheet

1. Find the exact values of each expression, where $0° \leq \theta < 360°$:

   a.) $arc \cos(\frac{-\sqrt{2}}{2}) =$

   $135°, 225°$

   b.) $arc \sin\left(\frac{-\sqrt{3}}{2}\right) =$

   $240° , 300°$

   c.) $\cos(arc \tan(-1)) =$

   $\pm\sqrt{2}/2$

   d.) $\sin(arc \tan(\sqrt{3})) =$

   $\pm\sqrt{3}/2$

   e.) $arc \cos\left(\sin\left(\frac{5\pi}{4}\right)\right) =$

   $135°, 225°$

   f.) $\cos(arc \sin\left(\frac{-\sqrt{3}}{2}\right)) =$

   $\pm 1/2$

2. Write the following as an algebraic expression in terms of x:

   a.) $\cot(arc \tan x) =$

   $\frac{1}{x}$

   b.) $\sin(arc \cos x) =$

   $\sqrt{(1-x^2)}$

   c.) $\sec(arc \csc \frac{2}{x}) =$ $\frac{2}{\sqrt{(4-x^2)}}$

PreCalculus Section 5.1 Work Sheet

1. Use the values of $\csc \theta = \frac{-13}{12}$ and $\cos \theta > 0$ to find the values of all six trigonometric functions.

    $\sin\theta = -12/13$ $\cos\theta = 5/13$ $\tan\theta = -12/5$

    $\csc\theta = -13/12$ $\sec\theta = 13/5$ $\cot\theta = -5/12$

2. If $\sec \theta = -\frac{5}{4}$ and $\csc \theta > 0$, then $\cot \theta =$

    $\sin\theta = \frac{3}{5}$ $\cos\theta = -\frac{4}{5}$ $\tan\theta = -\frac{3}{4}$

    $\csc\theta = \frac{5}{3}$ $\sec\theta = -\frac{5}{4}$ $\cot\theta = -\frac{4}{3}$

3. Simplify each of the following:

    a.) $2\sec^2 \theta - 2\tan^2 \theta = 2$

    b.) $\sec x - \sin x \tan x = \cos x$

    c.) $\frac{\csc^2 x - 1}{\csc x - 1} = \csc x + 1$

    d.) $\sec^3 x - \sec^2 x - \sec x + 1 = (\sec x - 1)(\tan^2 x)$

    e.) $2\sin^2 x - 18\cos^2 x = 2(1 - 10\cos^2 x)$

    f.) $6\sin^2 x + 3\sin x - 3 = 3(\sin x + 1)(2\sin x - 1)$

PreCalculus Section 5.2 Work Sheet

1. Simplify each of the following:

   a.) $\csc x \sin x =$ 1

   b.) $\sin x + \cot x \cos x =$ csc x

   c.) $3\tan^3 x - 192 \tan x$ 3 tanx (tanx-8) (tanx+8)

   d.) $\dfrac{3\sin x + 6}{4 - \sin^2 x} + \dfrac{4}{\sin x - 2}$ 1/(sinx-2)

2. Verify each of the following identities:

   a.) $\dfrac{\csc \theta}{\sec \theta} = \cot \theta$

   $= \cos\theta/\sin\theta$

   $= \cot\theta$

   b.) $\cos x + \sin x \tan x = \sec x$

   $= \cos x + \sin x \, \sin x/\cos x$

   $= (\cos^2 x + \sin^2 x)/\cos x$

   $= 1/\cos x = \sec x$

   c.) $\cos x(\sec x - \cos x) = \sin^2 x$

   $= 1 - \cos^2 x$

   $= \sin^2 x$

   d.) $\dfrac{\cot^2 x}{1 + \csc x} = \csc x - 1$

   $= (\csc^2 x - 1)/(1 + \csc x)$

   $= \csc x - 1$

PreCalculus Section 5.3 Work Sheet

1. Solve $\tan^2 x = \tan x + 2$ in the interval: $[0, 2\pi]$.

3π/4, 7π/4, 1.107, 4.249

2. Given: $2\cos x - 1 = 0$

   a.) Find the solutions in the interval: $[0, 2\pi)$.

π/3, 5π/3

   b.) Write the equation(s) for all the possible solutions:

π/3+2πn, 5π/3+2πn

(n= integer)

3. Solve $\cos x + \sqrt{2} = -\cos x$ in the interval: $[0, 2\pi)$.

3π/4, 5π/4

4. Solve $2\sin^2 x = 2 + \cos x$ in the interval: $[0, 2\pi)$.

π/4, 3π/2, 2π/3, 4π/3

5. Solve $\sin 2x = \dfrac{-\sqrt{3}}{2}$ in the interval: $[0, 2\pi)$.

2π/3, 5π/6, 5π/3, 11π/6

PreCalculus Section 5.4 Work Sheet

1. a.) Find the exact value of $\cos\frac{\pi}{12}$. $(\sqrt{6}+\sqrt{2})/4$

    b.) Find the exact value of $\sin 75°$. $(\sqrt{6}+\sqrt{2})/4$

    c.) Find the exact value of $\cot 375°$ $\sqrt{3}+2$

2. Simplify:

    a.) $\sin\theta\cos 2\theta + \cos\theta\sin 2\theta =$ $\sin(3\theta)$

    b.) $\cos 25°\cos 15° - \sin 25°\sin 15° =$ $\cos 40°$

3. If $\tan A = \dfrac{5}{12}$ and $\sin B = \dfrac{3}{5}$, where A and B are acute angles, find the value of $\cos(A + B)$.

    $\dfrac{33}{65}$

4. Find the exact value of $\tan 75°$.

    $\sqrt{3}+2$

5. Let $\cos\alpha = \dfrac{4}{5}$ and $\sin\beta = \dfrac{5}{13}$ where $\dfrac{3\pi}{2} < \alpha < 2\pi$ and $\dfrac{\pi}{2} < \beta < \pi$. Find $\sin(\alpha - \beta)$.

    $\dfrac{16}{65}$

PreCalculus Section 5.5 Work Sheet

1. Express $\sin 4x$ in terms of $\sin x$ and $\cos x$.

4sinx cos3x - 4sin3x cosx

2. Given that $\pi < \theta < \frac{\pi}{2}$ and $\cos \theta = \frac{-15}{17}$, find the value for each of the following:

   a.) $\sin 2\theta =$

$$\frac{-240}{289}$$

   b.) $\cos 2\theta =$

$$\frac{161}{289}$$

   c.) $\tan 2\theta$

$$\frac{-24}{161}$$

   d.) What quadrant is 2θ in?   quad. 4

3. If $\sin \theta = a$, find the value of $\sin 2\theta$ in terms of a.

$2a\sqrt{(1-a^2)}$

4. Using the half-angle formula, find the exact value of $\tan 75°$.

$\sqrt{3}+2$

5. Solve $\sin x \sin 40° - \cos x \cos 40° = \frac{1}{2}$ for all value of x such that $0° \leq x \leq 360°$.

80°, 200°

6. For what values of x between 0 and 2π is $\sin x < \cos x$?

$0<x<\frac{\pi}{4}$ or $\frac{5\pi}{4}<x<2\pi$

7. If $\cos 20°=b$, find the value of $\cos 40°$ in terms of b.

$2b^2-1$

PreCalculus Section 5 Chapter Review

1. Given $\sin\theta = \frac{1}{5}$. Find the $\cos\theta$ if θ is in the 2nd quadrant.

$-\dfrac{\sqrt{24}}{5} = -\dfrac{2\sqrt{6}}{5}$

Simplify the following:

2. $\sin x \tan x \cos x$

$\sin^2 x$

3. $\csc A \tan A \cos A$

1

4. $\dfrac{\sin X}{\cos X} + \dfrac{\cos X}{\sin X}$

$\csc x \sec x$

5. $\dfrac{\sin^2\theta - \cos^2\theta}{\sin\theta - \cos\theta}$

$\sin\theta + \cos\theta$

6. $\cot\theta + \dfrac{1 - 2\cos^2\theta}{\sin\theta\cos\theta}$

$\tan\theta$

7. Verify that $\cos^2\theta - \sin^2\theta = 1 - 2\sin^2\theta$.

$= \cos 2\theta$

$= 1 - 2\sin^2\theta$

8. Verify that $\dfrac{1}{1-\cos x} + \dfrac{1}{1+\cos x} = 2 + 2\cot^2 x$

$= 2/(1-\cos^2 x)$

9. Find the solutions in the interval $[0, 2\pi)$ of $\cos^2 x + \cos x - 2 = 0$.

$x = 0$

10. Find the solutions in the interval $[0,2\pi)$ of $2\cos x - 1 = 0$.

$\dfrac{\pi}{3}$, $\dfrac{5\pi}{3}$

11. Find the exact value of the $\cos 195°$.

$\dfrac{-(\sqrt{6} + \sqrt{2})}{4}$

12. Find the exact value of the $\tan 15°$.

$2 - \sqrt{3}$

13. Suppose $\sin\theta = \dfrac{4}{5}$ and $\dfrac{\pi}{2} < \theta < \pi$, find the $\sin 2\theta$, $\cos 2\theta$., and $\tan 2\theta$.

sin2θ= −24/25

cos2θ= −7/25

tan2θ= 24/7

14. Find the solutions on $[0,2\pi)$ of $\sin^2 x + 2\cos x = 2$.

x = 0

15. Find the exact value of the trigonometric functions given that the $\sin u = \dfrac{3}{4}$, $\cos v = \dfrac{-5}{13}$ and u and v are in Quadrant II.

    a.) $\sin(u + v) =$

       $= \dfrac{-(15+12\sqrt{7})}{52}$

    b.) $\cos(u - v) =$

       $= \dfrac{(36+5\sqrt{7})}{52}$

    c.) $\tan(2u) = 3\sqrt{7}$

PreCalculus Section 6.1 Work Sheet

1. Given the following values, determine how many triangles can be formed.

a.) a=15.2, b=8.5, and ∠B=42°

   No triangle

b.) c=28, b=11, and ∠B=40°

   No triangle

c.) b=8, a=7, and ∠B=30°

   2 triangles

2. Given triangle ABC with ∠A=39°, ∠B=106°, and c=78. Find a.

   85.58

3. Find all six parts of triangle ABC, given ∠A=48°, ∠C=57°, and b=47.

   ∠B= 75°,   a= 36.16,   c= 40.81

4. A cellular phone signal tower sits on the ground. Two 88-foot guy wires are positioned on opposite sides of the tower. The angle of elevation each wire makes with the ground is 21°. How far apart are the ends of the two guy wires?

   164.31ft

PreCalculus Section 6.2 Work Sheet

1. Find the number of degrees in the other two angles of $\triangle ABC$ if $c = 75\sqrt{2}$, $b = 150$, and $\angle C = 30°$

$\angle A = 20.705$ $\angle B = 129.295°$

2. In triangle ABC, a=9, b=13, and $\angle C=82°$. Find the length of c.

$c = 14.75$

3. In triangle ABC, a=13.2, b=18.5, and c=26.2. Find the largest angle.

$110.36°$

4. A tunnel is to be dug from point A to point B. The distance from a third point C to A is 3.65 miles and from point C to B is 2.74 miles, and $\angle ACB=49.2°$. How long will the tunnel be in length?

2.786mi

5. A surveyor finds the edges of a triangular lot to measure 5.3m, 10.5m, and 14m. Find the area of the lot.

$23.8m^2$

6. Find the area of $\triangle ABC$ if $a = 180$ inches, $b = 150$ inches, and $\angle C = 30°$

$6750 \ in^2$

PreCalculus Section 6.3 Work Sheet

1. Given the vectors a, b, c, and d sketch the following:

a.) 2a + c

b.) 3d − 2b

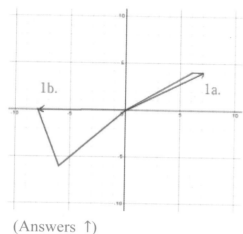

(Answers ↑)

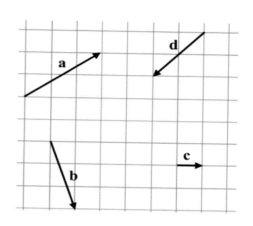

2. Given the vectors a, b, c, and d find the component of each vector (assume 1 unit = 1):

   a.)   2b   (-4, -4)

   b.) -2c+d (-4, -2)

   c.) 2a+c   (7,4)

3. A vector v has initial point (4,10) and terminal point (-4, -5). Find the following:

   a.) Component Form: (-8, -15)

   b.) $\|v\|$ $\sqrt{289}$= 17

   c.) Direction of the vector in component form: 241.93°

   d.) A unit vector u in the direction of v.   (-8/17, -15/17)

4. Let u = <4,5> and v = <8,-5>. Find each of the following:

   a.)   u − v =   (-4, 10)

   b.)   2u + 3v =   (32, -5)

   c.)   Write u as a linear combination of i and j. 4i+5j

PreCalculus Section 7.1 Work Sheet

1. Simplify: $a.)\ \dfrac{5!}{4!3!} =$

   a. 5/6

   $b.)\ \dfrac{2(n+2)!}{3n!} =$

   b. [2(n+2)(n+1)]/3

2. Write the first four terms of the sequence whose n^{th} term is $a_n = \dfrac{2n}{n^2+1}$.

   1, 4/5, 3/5, 8/17

3. Find a formula for the nth term of the sequence:

   -7, 2, 11, 20, 29, 38,…

   9n-16

4. Write the series in expanded form and evaluate:

   $$\sum_{i=2}^{6}(3i-1) =$$

   Sum= 55

5. Use sigma notation to write the sum:

   a.) $1 + \dfrac{1}{2} + \dfrac{1}{4} + \dfrac{1}{8} + \dfrac{1}{16}$

   Sigma Notation: $\sum_{i=1}^{5} \dfrac{1}{2^{i-1}}$

   b.) $2 + \dfrac{2}{2} + \dfrac{2}{5} + \dfrac{2}{10} + \dfrac{2}{17}$

   Sigma Notation: $\sum_{i=0}^{4} \dfrac{2}{i^2+1}$

6. Write the first four terms of the sequence defined recursively:

   $a_1 = 3$ and $a_{k+1} = 2a_k - 1$

   $3, 5, 9, 17$

PreCalculus Section 7.2 Work Sheet

1. Determine whether the sequence is arithmetic. If it is, find the common difference.

   a.) 3, 9, 15, 21, 27,....   arithmetic, d= 6

   b.) $1^2, 2^2, 3^2, 4^2, 5^2, ...$ Not arithmetic

   c.) $\frac{1}{2}, \frac{1}{4}, \frac{1}{6}, \frac{1}{8}, ...$ Not arithmetic

2. Write the first four terms of the arithmetic sequence:

   a.) $a_1 = 13$ and $a_{b+1} = a_b + 5$

   13, 18, 23, 28

3. Write the first four terms of the arithmetic sequence with $a_1 = 13$ and $d = -2$.

   13, 11, 9, 7

4. Find the 12th term of the arithmetic sequence with $a_1 = -23$ and d=-4.

   -67

5. Find the sum of the first 25 terms of the arithmetic sequence: 15, 25, 35, 45, 55,...

   3375

PreCalculus Section 7.3 Work Sheet

1. Write the first four terms of the geometric sequence with $a_1 = 6$ and $r = \frac{3}{2}$.

$6, 9, \frac{27}{2}, \frac{81}{4}$

2. Find the seventh term of the geometric sequence: $\frac{5}{64}, -\frac{5}{16}, \frac{5}{4}, ...$

320

3. Find the sum: $\sum_{k=2}^{7} 5^{k-2}$

3906

4. Find the sum of the infinite geometric series: $8 - 4 + 2 - 1 + \frac{1}{2}, ...$

$\frac{16}{3}$

5. Write the following sum in summation notation (Σ)

$$7 + 14 + 28 + ... + 896$$

$\sum_{i=0}^{8} 7(2)^{i-1}$

6. Evaluate (a.) $\sum_{n=1}^{\infty} \frac{3}{5^n} = $ 3/4

(b.) $\sum_{n=1}^{\infty} \frac{3^n}{5^n} = $ 3/2

PreCalculus Section 7.4 Work Sheet

1. Evaluate: 7C2 and 8P3: $21, 336$

2. Write the first three terms of the expansion and simplify: $(2x - 3)^6$

$(2x)^6+6(2x)^5(-3)+15(2x)^4(-3)^2+20(2x)^3(-3)^3...$

$= 64x^6 - 576\,x^5 + 2160x^4....$

3. Find the 5th term of $(a + 2b)^{10}$.

$_{10}C_4 a^6(2b)^4 ==> 3360\,a^6b^4$

4. Determine the coefficient of x^3y^y in the expansion of $(\frac{1}{4}x - 2y^2)^7$.

$\frac{35}{4} x^3y^8$

5. Expand and simplify: $(2 - 3b)^3$

$8-36x+54x^2-27x^3$

6. Give the first three terms of $(a + b)^{-3}$.

$\frac{1}{b^3} - \frac{3b}{a^4} + \frac{6b^2}{a^5}$

PreCalculus Section 8.1 Work Sheet

1. Given: $x^2 = 12y + 1$.　　　Vertex: $(0, -\frac{1}{12})$　　Focus: $(0, \frac{35}{12})$

　　　　　　　　　　　　　　　　Directrix Line:　　　$y = -\frac{37}{12}$

　　　　　　　　　　　　　　　　Does it go up, down, left, or right? Up

2. Given: $y^2 = -8x$　　　　　Vertex: $(0,0)$　　Focus: $(-2,0)$

　　　　　　　　　　　　　　　　Directrix Line:　　　$x = 2$

　　　　　　　　　　　　　　　　Does it go up, down, left, or right? Left

3. Graph the following: $y^2 + 2y - x + 1 = 0$. Label the vertex, focus, directrix, and axis of symmetry.

vertex $(0, -1)$

focus $(\frac{1}{4}, -1)$

Axis of Sym.$(y = -1)$

directrix $x = -\frac{1}{4}$

4. Write the standard form of the equation of the parabola with Vertex: (-1,2) and Focus: (-1,0)

$(x+1)^2 = -8(y-2)$

5. Write the standard form of the equation of the parabola with Vertex:(2,4) and directrix line: (x = 4).　　$(y-4)^2 = -8(x-2)$

PreCalculus Section 8.2 Work Sheet

1. Given: $\frac{(x+2)^2}{16} + \frac{(y-3)^2}{9} = 1$. Find the following and make a sketch:

Sketch

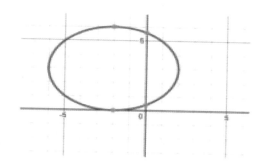

a.) Center: (-2, 3)

b.) Vertices: (2,3), (-6,3)

c.) Endpoints of Minor Axis: (-2,6), (-2,0)

d.) Length of Major Axis: 8

e.) Length of Minor Axis: 6

f.) Eccentricity= $\sqrt{7}/4$

g.) Foci: $(-2, 3 \pm \sqrt{7})$

2. Write the following in standard form and sketch. Label Center, vertices, and foci

$$x^2 + 16y^2 - 160y + 384 = 0.$$

center: (0,5)

vertices: (4,5), (-4,5)

foci: $(\pm\sqrt{15}, 5)$

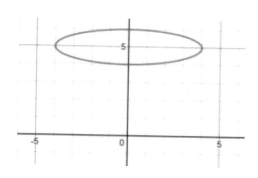

3. Write the equation of the ellipse in standard form with Foci $(0, \pm 1)$, and Vertices $(0, \pm 4)$.

$x^2/15 + y^2/16 = 1$

PreCalculus Section 8.3 Work Sheet

Graph the below equation and indicate all important points.

1. $y = 5x^2 - 10x + 6$

2. $x^2 - 4y^2 + 2x + 8y = 7$

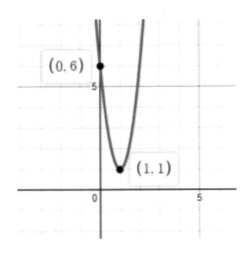

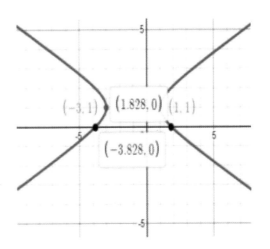

3. $16x^2 + 25y^2 - 32x - 150y = 159$

4. $x^2 + y^2 + 8x = -12$

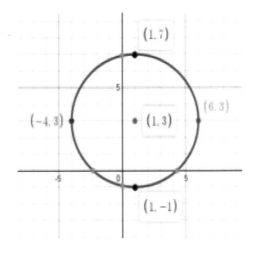

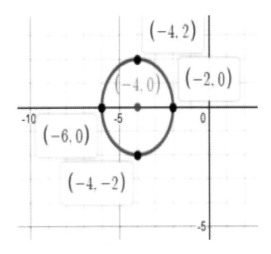

PreCalculus Section 8.4 Work Sheet

1. Sketch the graph of $\frac{(y-2)^2}{25} - \frac{(x+3)^2}{16} = 1$.

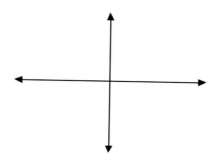

Center= $(-3,2)$

Focus Points: $(-3, 2\pm\sqrt{41})$

Vertices: $(-3,7)$, $(-3,-3)$

Length of Transverse Axis: 10

Equations of the asymptotes: $(y-2)= \pm\frac{5}{4}(x+3)$

2. Sketch the graph of $x^2 - 4y^2 + 2x + 8y = 7$. Label Center, vertices, and foci.

center $(-1,1)$

vertex $(1,1)$, $(-3,1)$

focus $(-1\pm\sqrt{5},1)$

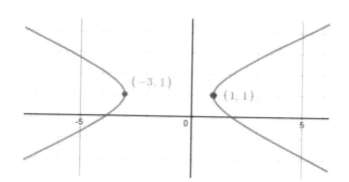

3. Write the equation of the hyperbola in standard form with Center $(-1, 5)$; Vertex $(-1, 6)$; and Focus $(-1, 9)$

$(y-5)^2-(x+1)^2/15= 1$

PreCalculus Section 8.5 Work Sheet

1. Sketch the curve represented by the parametric equation. Then, eliminate the parameter and write the corresponding rectangular equation whose graph represents the curve. Indicate the direction of the curve.

   a.) $x = 1 - 3t$ and $y = 5 + 2t$

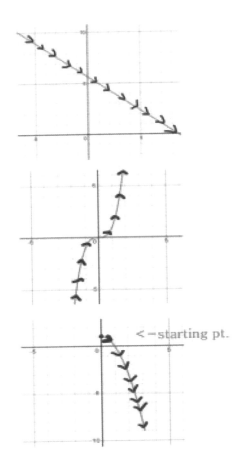

   b.) $x = t$ and $y = t^3$

   c.) $x = \sqrt{t}$ and $y = 1 - t$

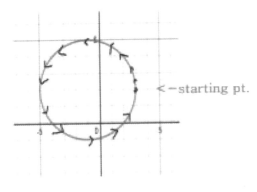

   d.) $x = 4\cos t - 1$ and $y = 3\sin t + 2$

PreCalculus Section 8.6 Work Sheet

1. Plot the points and give three additional polar representations.

   a.) $\left(2, -\frac{7\pi}{4}\right)$

   b.) $\left(-3, \frac{5\pi}{6}\right)$

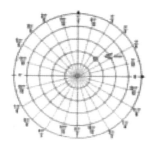

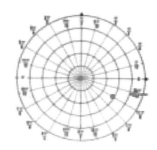

many possible answers!!

Additional Representations

(-2, -3π/4) or (2, -π/4)

Additional Representations

(3,-π/6), or (3, 11π/6)

2. Convert $(5\sqrt{2}, -\frac{11\pi}{6})$ to rectangular form.

$((5\sqrt{6})/2, (5\sqrt{2})/2)$

3. Convert (3, -1) to polar form.

$(\sqrt{10}, -18.43°)$

4. Convert the rectangular equation $x^2 + y^2 - 8y = 0$ to Polar form.

$r = 8\sin\theta$

5. Without sketching, describe the graphs of (a.) $r = 2$ and (b.) $r = \frac{1}{\sin\theta}$.

a.) =>circle at the origin with $r = 2$

b.) = > line: y= 1

PreCalculus Section 8.7 Work Sheet

1. Sketch the graph of the polar equations. Fill out the table as well.

a.) $r = 4\cos\theta$

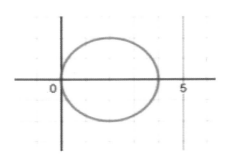

θ (degrees)	0	30°	60°	90°	120°	150°	180°	210°	270°	330°	360°
radians	0	$\frac{\pi}{6}$	$\frac{\pi}{3}$	$\frac{\pi}{2}$	$\frac{2\pi}{3}$	$\frac{5\pi}{6}$	π	$\frac{7\pi}{6}$	$\frac{3\pi}{2}$	$\frac{11\pi}{6}$	2π
r	3	2.6	1.5	0	-1.5	-2.6	-3	-2.6	0	2.6	3

b.) $r = 5 - 4\sin\theta$

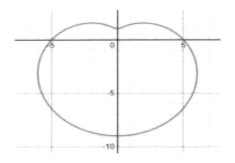

θ (degrees)	0	30°	60°	90°	120°	150°	180°	210°	270°	330°	360°
radians	0	$\frac{\pi}{6}$	$\frac{\pi}{3}$	$\frac{\pi}{2}$	$\frac{2\pi}{3}$	$\frac{5\pi}{6}$	π	$\frac{7\pi}{6}$	$\frac{3\pi}{2}$	$\frac{11\pi}{6}$	2π
r	5	3	1.54	1	1.54	3	5	7	9	7	5

PreCalculus Conic section review questions

1. Each of the following is an equation of a conic section. State which one and find, if they exist:

(I) the coordinates of the center, (II) the coordinates of the vertices, (III) the coordinates of the foci, (IV) the eccentricity, (V) the equations of the asymptotes. Also, sketch the graph.

(A) $9x^2 - 16y^2 - 18x + 96y + 9 = 0$

Hyperbola.
I. center (1,3)
II. vertices: (1,6), (1,0)
III. foci (1,8), (1,-2) $f_L = 5$
IV. $e = \frac{5}{3}$
V. $y-3 = \pm\frac{3}{4}(x-1)$

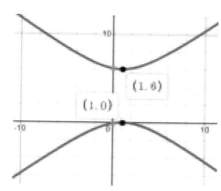

(B) $4x^2 + 4y^2 - 12x - 20y - 2 = 0$

circle.
I. center $(\frac{3}{2}, \frac{5}{2})$, r= 3

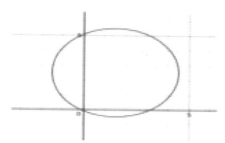

(C) $4x^2 + y^2 + 24x - 16y = 0$

ellipse
I. center (-3,8)
II. vertices: (-3,18), (-3,-2)
III. foci (-3,8±5√3) $f_L = 5\sqrt{3}$
IV. $e = \sqrt{3}/2$
V. No. Asymp.

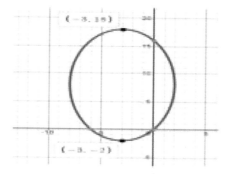

(D) $y^2 + 6x - 8y + 4 = 0$
Parabola
I. No. center
II. vertix: (2,4)
III. focus (1/2, 4) $f_L = -3/2$
IV. No e
V. No. Asymp

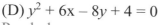

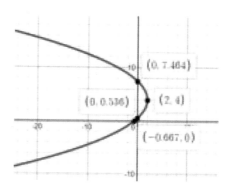

2. Find the equation of the hyperbola with center at (3, -4), eccentricity 4, and transverse axis on y axis of length 6.

$$\frac{15(y+4)^2}{9} - \frac{(x-3)^2}{9} = 1$$

3. What is the graph of each of the following?

(A.) $x^2 + y^2 - 4x + 2y + 5 = 0$ circle

(B.) $xy = 0$ 2 intersecting lines

(C.) $2x^2 - 3y^2 + 8x + 6y + 5 = 0$ 2 intersecting lines

(D.) $3x^2 + 4y^2 - 6x - 16y + 19 = 0$ points

(E.) $x^2 + y^2 + 5 = 0$ Do not exist. Empty set

PreCalculus Section 9.1 Work Sheet

1. Find the limit of $\lim\limits_{x \to 4}(x^2 - 3x + 1)$ by using the table feature on the calculator.

x	3.9	3.99	3.999	4	4.001	4.01	4.1
f(x)	4.51	4.950	4.9950	5	5.005	5.05	5.51

Limit of f(x) as x approaches 4 is   5

2. Find the limit of $\lim\limits_{x \to 1}\dfrac{x - 3}{x^2 - 4x + 3}$ by using your graphing calculator and graphing it.

Limit of f(x) as x approaches 1 is   -1

3. Does the limit exist why or why not.

a.) $\lim\limits_{x \to 3}\dfrac{|x-3|}{x-3}$ No limit

b.) $\lim\limits_{x \to 0}\dfrac{2e^x-2}{x-2}$ exist = 0

c.) $\lim\limits_{x \to 0}\dfrac{\sqrt{x+5}-4}{x-2}$ exist $(\sqrt{5}-4)/-2$

4. Find the limits:

a.) $\lim\limits_{x \to -2}(2x^3 - 6x + 5)$ b.) $\lim\limits_{x \to -5}\dfrac{8}{x + 2}$ c.) $\lim\limits_{x \to 8}\dfrac{\sqrt{x + 1}}{x - 2}$

   1                    $\dfrac{-8}{3}$ $\dfrac{1}{2}$

1. Evaluate the limits:

a.) $\lim\limits_{x \to -3} (\frac{1}{3}x^2 - 4x)$ 15

b.) $\lim\limits_{x \to 4} \dfrac{x - 2}{x^2 + 2x + 2}$ $\dfrac{1}{13}$

c.) $\lim\limits_{x \to 3} \dfrac{x^2 - 5}{3x}$ $\dfrac{4}{9}$

2. Find the limits, if they exist.

a.) $\lim\limits_{x \to 4} \dfrac{4 - x}{x^2 - 16}$ $\dfrac{-1}{8}$

b.) $\lim\limits_{a \to -3} \dfrac{a^3 + 27}{a + 3}$ 27

c.) $\lim\limits_{b \to 0} \dfrac{\sqrt{7 - b} - \sqrt{7}}{b}$ $\dfrac{-\sqrt{7}}{14}$

3. Approximate the limit to three decimal places:

a.) $\lim\limits_{x \to 0} \dfrac{\sin x}{2x}$ $\dfrac{1}{2}$

PreCalculus Section 9.3 Work Sheet

1. Use the definition of Derivative to find the slope of the graph of the function at the specified point.

a.) $h(x) = 3x + 5$ at $(-1, -3)$ 3

b.) $f(x) = 2x - x^2$ at $(3, 12)$ -4

c.) $f(x) = \frac{2}{(x-1)}$ at $(2, 1)$ -2

2. Find the derivative of $f(x) = x^3 + 3x$. Determine where the tangent line is horizontal.

$f\ '(x) = 3x^2 + 3$, so there is no horizontal tangent line.

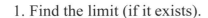

PreCalculus Section 9.4 Work Sheet

1. Find the limit (if it exists).

a.) $\lim\limits_{x \to \infty} \dfrac{3}{2 + x}$ 0

b.) $\lim\limits_{x \to \infty} \dfrac{5 - 2x}{1 + 3x}$ $-2/3$

c.) $\lim\limits_{y \to \infty} \dfrac{3y^3}{y^2 - 32}$ 3

2, Find the limits of the sequences (if it exists.) Assume n begins with 1.

a.) $a_n = \dfrac{4n - 1}{n + 3}$ 4

b.) $a_n = \dfrac{4n - 5}{3}$ ∞

c.) $a_n = \dfrac{(-1)^{n+1}}{n^2}$ 0

Geometry

Answer Sheet

Geometry Review

Provides problems for review and self-diagnosis through appropriate review questions to see if you have a good understanding of geometry.

Triangle Review

1. D 2. C 3. D 4. E 5. D 6. D 7. C 8. B 9. B 10. D

Quadrilaterals and Other Polygons Review

1. B 2. E 3. C 4. D 5. C 6. E 7. C 8. A 9. B 10. A

Circle Review

1. D 2. D 3. D 4. B 5. E 6. E 7. D 8. A 9. A 10. C

Coordinate Geometry

1. B 2. D 3. A 4. C 5. B 6. B 7. A 8. E 9 B 10. D

Algebra2

Answer Sheet

Algebra2 Review

We provide questions for review and self-diagnosis through appropriate review questions to see if you have a good understanding of Algebra.

Practice Review Packet #1

1. D 2. C 3. E 4. E 5. A 6. A 7. D 8. A 9. A 10. A

11. A 12. A 13. B 14. A 15. E

Practice Review Packet #2

1. D 2. D 3. B 4. C 5. C 6. B 7. C 8. C 9. D 10. B

11. A 12. E 13. E 14. B 15. C

Complete Guide

to High School Math

PRECALCULUS

MASTER

GEOMETRY
ALGEBRA

REVIEW